The Patrick Moore Practical Astronomy Series

More information about this series at http://www.springer.com/series/3192

Rare Astronomical Sights and Sounds

Jonathan Powell

 Springer

Jonathan Powell
Ebbw Vale, United Kingdom

ISSN 1431-9756   ISSN 2197-6562   (electronic)
The Patrick Moore Practical Astronomy Series
ISBN 978-3-319-97700-3   ISBN 978-3-319-97701-0   (eBook)
https://doi.org/10.1007/978-3-319-97701-0

Library of Congress Control Number: 2018953700

This Springer imprint is published by the registered company Springer Nature Switzerland AG
The registered company address is: Gewerbestrasse 11, 6330 Cham, Switzerland

Contents

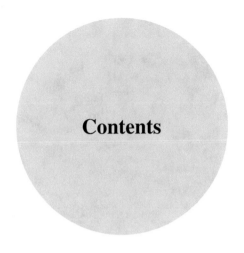

Chapter 1

Introduction: From the Past, into the Present, and onto the Future

Our ancestors lived and died by observations and rituals created around annual events in the sky. Such happenings gave a pattern of stability to their world: when to sow crops, when to reap crops, the onset of warmer temperatures, and the onset of colder temperatures, reflected by the seasons.

To acknowledge the sky and its perceived power, many rituals and sacrifices were made, all of which were observed with great timing and accuracy. Various civilizations had their own ways of respecting the rising of the Sun, the passage of the stars, and the phases of the Moon. In a comforting form of naivety, not fully understanding the mechanisms and drivers behind what they were witnessing, they still registered the impact the sky could have on their daily lives, an unquestioned authority that in our modern world has sadly been neglected. We must learn today as did our ancestors that while we can see and project our understanding of what we behold onto the sky, such interpretations are not as certain as we might think. Let us not forget that the way our ancestors saw the skies was continually challenged, corrected, and in some cases augmented over the thousands of years. Who is to say that our definition is ultimately the correct one?

Amongst the annual sights in ancient sky were rarities, oddities, and occurrences that would have proved very disturbing to the onlooker, as such events broke with tradition, upset the equilibrium, and penetrated that comforting bubble of naivety. Eclipses in particular humbled and frightened those who witnessed them. Whilst in modern times, we can explain why and

© Springer Nature Switzerland AG 2018
J. Powell, *Rare Astronomical Sights and Sounds*, The Patrick Moore
Practical Astronomy Series, https://doi.org/10.1007/978-3-319-97701-0_1

how they occur, they are still relatively rare and continue to produce a sense of wonder. Perhaps they are more common than once-in-a-lifetime events, but they are still rare when speaking of the number of people who have actually witnessed a total solar eclipse, or for that matter, a transit of Mercury, thunderclouds on Saturn, or closer to home, a moonbow, a sun pillar, a green flash.

Everyday Life

Many people from many walks of life seem to gravitate towards a plateau-like zone of existence, a state of being that affords a level of knowing, understanding, and appreciation for what came before and what lies ahead. It may well take weeks, months, or years for an individual to finally enter this stable state. The feat may for a majority prove satisfying well beyond the initial moment of achievement, instilling them with a sense of fulfillment that makes them want to remain within that plateau. Maintaining this level of comfort can become a priority that subsequently outweighs ambitions to seek other opportunities or paths leading out of that zone. For the minority, the satisfaction will always be fleeting, prompting the questions: what next, where next, and how to achieve it?

Those who have achieved existence in their own plateau-like zone (often unique to the individual in its look, feel, and attributes) may well have entered a place that they argue cannot be bettered, or the risk of potentially bettering what they have is too great. Others argue that in order to realize other zones, further sacrifice is required and, in reality, no one zone will ever offer exactly the right ingredients that would allow for the journey to finally end.

These zones are not life goals driven by results, performance, and expectations set by others, but zones that exist within everyday life—from the moment one wakes to the moment one goes back to sleep. What distinguishes the passage of time to and from multiple zones in one's life are certain key features that turn ordinary to extraordinary, where the automation is temporarily eclipsed by a bright flash, or a sound momentarily breaks the silence in a world that has so much noise.

These flashes of light or infrequently heard sounds represent the rarities in everyday life, the reality checks that, regardless of what stage we have reached in life, have the power to impact our whole outlook and understanding of the larger picture that exists for us all. A simple action, word, or deed can have a similar affect, becoming a turning point or junction that alters a course or changes an opinion. How many times has it been said that if we could hold one precious, singular moment in life and make that moment resonate for the rest of our lives, then there would exist a true sense of fulfillment that cannot be bought, earned, or bartered for? A newborn child held in a mother's arms represents the cycle of life at its very beginning, a moment that for both are frozen in time, held in the moment, and cherished for the future.

These life rarities can be applied not just on a daily basis and not just on one solitary aspect of it, but also to a much larger and more grandiose scale. Ultimately, we all exist on the same level within our Universe, and the workings and patterns of our everyday life form part of a significantly more complex picture. It is this search for the big picture that gifts those with enough passion the ability to see the intricate and often unexperienced rarities offered by our awe-inspiring cosmological world. This world acts as a timepiece that not only observes our own time on Earth, but also the general passage of everything that has been, everything that was, and when "the time" comes, everything that will eventually be.

Fig. 1 The expanding universe. Courtesy of NASA

Universal Timepiece

Astronomical Rarities delves into the horological side of the universe, looking at some of the finer workings of the timepiece that creates its own flashes and sounds. Whilst rare, these events are as important in substance and nature as the most prolific and larger aspects in the universe, as one cannot exist without the other.

A river can be crossed many times, but it does not always have to be done at the same crossing point. The journey is different and unique to each person, and its undertaking does not require a specific commitment level, just a desire and a passion to see and hear some of the lesser workings of the cosmological clock. The unity between our everyday existence and the universe is as far-reaching and important as words can express, for we too are part of the timepiece, part of the mechanism, and part of one of the greatest adventures of all.

The mere recognition of our place in the universe was an aspect of life that our ancestors quickly recognized, a part of their culture that was in many ways much more rooted in their everyday lives than in modern times, back when the people, the land, and the skies had a closer bond with the universe. They looked upon the skies as they would look upon their own offspring, with a realization of the great bond existing between them and their surroundings. To take this link for granted would be to commit sacrilege.

This book is full of the observable rarities in our Universe, those that we know and understand, or at least partially understand. There remains a great deal that we do not understand, and a good amount of astronomical events that must occur for other events to become visible to us often go unseen. In this sense, such rarities are like an answer key: the end result may be evident, but how it was reached is what's important.

A great example of an astronomical event that we know regularly occurs, but that in the vast number of cases we do not observe, is a meteor strike, not just on Earth but on other planets as well. Our awareness of such strikes has been heightened over the years thanks to advances in technology, but even with these developments, space still holds the element of surprise.

On Saturday June 2, 2018, astronomers working at the Catalina Sky Survey in Arizona discovered a small asteroid near the orbit of the Moon, subsequently labeled 2018 LA. Several hours later, 2018 LA, a boulder-sized space rock traveling at 17 kilometers a second, struck the Earth, exploding over Botswana at 6.44 pm local time. A video camera at a farm near Ottosda, South Africa recorded the event. The explosion was not only visual, but also sent waves of low-frequency sound (infrasound) rippling through the Earth's atmosphere. These waves were detected at Station 147, an infrasound monitor in South Africa. The two-meter-diameter asteroid caused strong infrasound, with a yield range of 0.3 to 0.5 kilotons of TNT. While the atmosphere was able to burn much of 2018 LA into cinders, fragments were discovered across its path.

Warning alerts for 2018 LA went out less than a day before it struck. Hopefully, larger bodies will give lead time to observe their trajectory. As it stands, observation of an incoming projectile is about as much as we can do—an experience not dissimilar to that of the dinosaurs.

Chapter 2

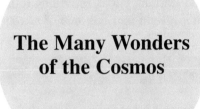

The Many Wonders of the Cosmos

Let's See What's Out There!

There is much to be seen and to behold in the universe. Many of these wonders are within easy reach of the astronomer, amateur and professional alike. We pride ourselves on having pushed back the boundaries on Earth with the great explorers like Marco Polo, Ferdinand Magellan, and Captain Cook setting sail to see exactly what our planet had to offer in terms of new lands and new civilizations. And yet, despite all the explorers of the past and present, just 5% of the Earth's oceans has been explored. Considering the oceans make for 70% of the Earth's surface, we still have much work to do right in our own backyard.

Humankind's restlessness has had us reaching for the stars long before we ever made the smallest progress in unraveling the mysteries of life right here on Earth. As of 2010, biologists have described and classified 1.7 million plants and animals, less than one quarter of the total species estimated in the world. Scientists predict there are still over five million species waiting to be found. Each year, scientists record 18,000 new species of plants and animals, with a study in 2012 claiming the average time between the discovery of a new species and its description is 21 years. After many years of describing, naming, and cataloguing the species we share our planet with, we remain a long way from gaining a complete picture. And yet, the desire to find out if we are alone in the universe remains an incredible driver for the exploration of space.

© Springer Nature Switzerland AG 2018

J. Powell, *Rare Astronomical Sights and Sounds*, The Patrick Moore
Practical Astronomy Series, https://doi.org/10.1007/978-3-319-97701-0_2

There are many wonders of the cosmos that are readily accessible to the amateur through ever advancing technology. Hearing and seeing as much of what space has to offer as possible requires a little bit of preparation and legwork. Before that journey can begin, we must consider other aspects that at first may appear to have only a tenuous link to the material, but are in fact a necessary part of the learning curve, providing tools that will ultimately allow one to gain better insight into this area of astronomy.

Our Place in the Universe: Time

Picture an antique clock on a mantelpiece. A clock that with its hands commands a quarterly hour chime, with a metronome-like *tick-tock* marking the passage of each second, minute, hour, and day. In the room where the clock is kept, there passes varying levels of human activity, all observed by the clock whose sole purpose is to mark time.

For some of the owners of the clock, the passage of time is rapid, the movement of the hands reflecting a swift dismissal of what has been, with an attempt to organize and ensure that what time remains in a given day is put to good use. The desire to master time and seize the day makes for an efficient and effective usage of the hours. "Look at the time! Where did the time go!?" remarks the one owner, horrified that time has marched past, and yet the fulfillment needed to make that time count has been lost. That time cannot be recovered; it has been consumed by the realm that expands with every second: the domain of history. Still, the clock on the mantelpiece ticks.

For another keeper of the clock, the passage of time is not so rapid, reflecting a steady but constant marking of minutes and hours throughout a day, often a day without stimuli or involvement with others or indeed engagement and interaction with the world around that person. "Time hangs heavy, it really is dragging today!" complains this other owner. This owner is aware of every fluctuation in the length of light in the day, where shorter days mark an earlier closure of the summer window, thus making the more sedate observer of time mark its passage with even greater apprehension. And still the clock on the mantelpiece ticks.

One could easily assume that the owner who strives to make the most of time is young, driven by an enthusiasm to see and do as much as they can before time makes them unable to do so. One could also say that the owner who marks time with a slow, regretful glance is older, possibly much older, at the point where time is running out. Whereas time may move rapidly along for one and drag on for the other, it is in fact precious for them both, as inevitably, their timelines will cease. And still the clock on the mantelpiece ticks.

Objective time moves at the same pace. It does not judge or react differently—it is merely the perception of time and its usage that makes the difference, with the clock on the mantelpiece the companion and observer to it all. We know what time is, we know how to mark its passage, but we can't say exactly what happens when time passes—it just does.

According to the general theory of relativity, the universe emerged following the Big Bang, with all matter before that concentrated down into an extremely tiny dot—ironically, a dot that can be paralleled with Carl Sagan's "Pale Blue Dot," as its complexity and existence are greatly mirrored. A tiny, tiny portion of this dot, which contained all matter, later became the Sun, the Earth, the Moon, and the Solar System, all of which when observed tell us all about the passing of time. Motion is the key. The motion of matter and the observation of time through motion will eventually, according to the general theory of relativity, lead to a point when the universe collapses. All matter would then shrink into a tiny dot once more, which would bring time to a halt.

Some of the greatest unanswered questions are linked to time. How does time flow, and does it flow in one direction only? Is time a universal constant? Although the measuring of time has changed in history with the introduction of new technology and new units and measurements, time itself has a remained a relative constant.

Physicists define time as the progression of events from the past to the present into the future. If a system is unchanging, it is timeless. Time is something that we can neither see nor touch, but, by motion, we can measure its passage. Equations relating to time work equally well whether time is moving forward into the future (positive time) or working backwards into the past (negative time), which is something to be considered when challenging the given premise that time moves forward and only forward in a constant, one-directional flow. This positive advancement is known as the "arrow of time," dictating that time cannot move backwards at all, and that it is impossible for the universe to return to any point on its timeline between the present and its long history, stretching back to that initial first dot.

We know from Einstein's special and general theory that time is relative, dependent on the frame of reference of an observer. Different viewpoints of time can result in time dilation. Time dilation is a phenomenon where the time between events becomes longer (dilated) the closer one travels to the speed of light. A moving clock runs more slowly than a stationary clock, with the effect becoming more pronounced as the moving clock approaches light speed. Clocks in orbit record time more slowly than those on Earth.

In your mind's eye, picture an ordinary sweep hand alarm clock moving at a relativistic speed (a speed comparable to the speed of light) in an orbit around the Earth. Back on Earth, an observer is able to see the face of the clock while also being able to take note of the time on another sweep hand alarm clock in their possession. As the clock travels in orbit, a rather bizarre occurrence is witnessed by the observer back on the ground. The observer, watching in an inertial reference frame as opposed to the moving clock, sees the moving clock slow down, progressively "losing time" against the observer's ground-based clock. The observer is presented with two different times: that of the clock in orbit, and that of the ground-based clock. So, which time is correct? Surely, as the clock that is in orbit has seen a slowing in time, the ground-based clock must be gauging the correct passage of time? In fact, both clocks are correct, because time is not absolute but relative: it depends on the reference frame.

Let us take another famous example and imagine that two people of the same age work for NASA; these two people are twins, born at exactly the same time. The first twin has a ground-based position in mission control, the other is an astronaut who is about to be sent on a deep space mission. At the time of departure, the clocks at mission control and onboard the deep space capsule read the same. The astronaut then embarks on the mission, whereby she will be traveling at 95% the speed of light. Upon returning from the mission, the astronaut's onboard clock has measured that 10 years have elapsed since she left Earth, aging the astronaut 10 years. However, after landing on Earth and meeting up with her twin, the astronaut sees that for her twin, 32 years have elapsed, not 10! How has this happened? This great difference in aging and the associated measuring of time is explained by the fact that the astronaut is traveling at relativistic speeds, therefore her clock has slowed down.

In the dimension of space, you can move both forward and backward, but with time, we are always moving forward. Setting science aside, ontology, a branch of metaphysics that studies the nature of existence or being, poses a question about the sort of ontological differences there are among the present, the past, and the future. Three competing theories exist that try to answer this conundrum.

First is the view of the presentist philosophy—that of a person who possesses an uncritical adherence to present-day attitudes, interpreting history in terms of modern values and concepts. The presentist argues that necessarily, only present objects and present experiences are real, and that our conscious state recognizes this in the special vividness of our present experience, compared to our dim memories of past experiences and our expectation of future experiences. For example, the extinction of the dinosaurs has slipped out of reality, even though our current ideas of them have not.

Second is the growing-past theory, which states that the past and the present are both real, but the future is not real, because the future is indeterminate or merely potential. The kingdom of the dinosaurs was real, but our future death is not.

The final theory is "externalism." This states that there are no objective ontological differences between the past, present, or future, because the differences between the three are merely subjective.

Whether or not we have been accurately measuring time or have been giving time less or more credence that it deserves remains a debatable issue, along with a whole plethora of questions. Much like fish cannot understand water because they are contained within it, we still have a very poor grip on time, the most immersive of all substances.

The Beginning?

To "start at the beginning" already makes an awful lot of assumptions. For one, it assumes that before the beginning there was nothing, and everything that we now know simply came into being at that time. This is a staggering piece of naivety that rightfully has never been fully accepted as law, as turning a blind eye to possible happenings before the Big Bang narrows the ways we might determine and understand what we now take as the universe.

Let us look for example at a group of walkers hiking in a national park. The path that they are on is unfamiliar; in fact, they are pioneering this route so that others in the years to come can follow in their footsteps. Should problems be encountered along the route, the walkers can simply make a note of something that should be avoided, or hint that there is a certain way for the problem to be overcome, or point out that whilst it is an issue for them, someone following behind may have a better insight into how to deal with it.

Some way along this route, another walker joins the stroll by random chance, with no planned intervention. This new element, which simply picks up the thread halfway through the route, has no idea where the journey started or where it will end.

In this analogy, the humans of the present are that random walker, and the group is their ancestors. In order for us present-day humans to truly understand how we came to be in this circle on this very path, we must determine: at what exact point on the route we have joined forces; where the walkers' start point was (the proliferation of our species); what other well-worn trails had been carved out prior to this group making its appearance (the introduction of all life on Earth); and what the national park looked like before any of the trails existed (the commencement of the cosmos–"first light").

Following the Big Bang, the universe was teeming with hydrogen atoms whirling around in the all-encompassing darkness. As millions of years passed, these hydrogen atoms began to form clouds. Over time, the massing of atoms produced larger and larger clusters until a critical point was reached, and through the process of nuclear fusion, light pierced the darkness for the first time. The first stars were born, and the Cosmic Dawn had finally brought the endless blackout to an end.

As the light from these first stars shone, the hydrogen gas that surrounded them became stimulated, priming the gas to absorb some of the leftover energy from the Big Bang. Some of this "cosmic microwave background" seemed to be tweaked to a very specific frequency, and if astronomers could locate this signal, it would pinpoint the period in the history of the cosmos where reionization commenced—a critical point on the route.

The Thirst for Knowledge Begins

The pursuit of answers to questions in all fields, not just science, can be painstaking, with years of dedication that ultimately might not bear any fruit at all. However, it is that possibility of failure that makes for an excellent personal driver, continually fueling ambition until all possible avenues are closed. The polarizing power of such positive and negative thoughts, along with the human instinct to not easily concede, has fueled continuous discoveries.

Despite the lack of technology available to early astronomers, countless maps were still drawn, observations were still made, and theories still proposed. As the centuries passed and technology developed, observational boundaries were pushed back, more detailed depictions were drawn, and long-untested theories soon found themselves either supported at last or quashed for good. Technology, properly acknowledged and correctly used, provides a necessary crutch that enhances our existence. It should not simply make it easier for us. Making it easier only takes away the desire to improve, expand, and discover. If as a race we are to survive, we must find a more natural balance that allows our human nature and our technology to coexist.

Once the technology that we've created is safely within its given boundaries, the conquest of space can be accelerated, with advancement and understanding delivered in faster timeframes. Already, many questions have been answered through technology and the fields of astronomy and space sciences. And yet our understanding, while increasing with every piece of data analyzed, only fills but a few percent of the sum total, a total which oftentimes only seems to grow larger. While that may sound exasperating, it is the very driver that keeps us pushing on and discovering greater things.

The aim of this book is to discover, examine, and ultimately explore the rarer sights that the universe has to offer. But, in keeping with our metaphors and theme about the importance of time and chronology, we will first need to examine all the moving parts. This is necessary so that progress beyond the point of arrival can be continued. If one makes a train journey that has a start point of A and an arrival point of K, one automatically presumes that the train had to pass through points B to J. However, if for some there were track repairs between D and E, it may have been necessary to involve a connecting bus in order to abridge the point where the train was unable to continue its journey. One minor change could alter the timetable immensely, causing or being caused by other problems and delays en route.

This concept of the past and past experiences applies most firmly to seeking out the rarities the cosmos has to offer, as the term "rarities" is relative to exactly how much one already knows. Our forefathers' steppingstone into the universe rested on only a basic understanding of the rudimentary workings of the world—forget about deep sky objects or radiation belts of faraway worlds!

At first, the wonders of the cosmos were limited to observing the heavens at night with the naked eye, beholding a sky not distorted by light pollution, along with meteor displays, passing comets, and our Moon. These phenomena would have meant different things to different belief and cultural systems. The Moon, viewed from anywhere on Earth, would most certainly have been the astronomical companion upon which most eyes rested, for its seeming omnipresence could provide a sense of continuity to a seemingly disorganized night sky. As the centuries have passed and understanding gained, the wonderment has only expanded as we probe deeper and deeper into space and, ironically, back in time!

As we seek out the rarities of today, we must be mindful of the rarities of the past. Some were rarities then as they are today, and while the capabilities of finding and observing such events has in some cases been greatly shortened, they still hold much appeal to those continuously searching for astronomy's precious gems.

Rocks That Preserve Our Past

The earliest evidence of life on Earth lies within some of the oldest rocks found on our planet.

In Australia, a set of filament-like fossils were reported in the journal *Astrobiology* in 2013. The contents of these fossils, dated back some 3.5 billion years, show evidence of microbial communities. In this era's ecosystem, these communities clung to sediments along a shoreline, surviving on the energy supplied by the Sun.

In Greenland, a set of rocks dating further back to 3.7 billion years show signs of ancient colonies of cyanobacteria. Tiny ripple or wave-like sediment formations measuring 0.4–1.5 inches resided on the ancient seafloor and were subsequently captured, housed, and held within these rocks. Known as the Isua Greenstone Belt in southwestern Greenland, these ripples are fossilized remains of cone-shaped "stromatolites," layered mounds of sediment and carbonates that build up around colonies of microbes that grow on the floor of shallow seas or lakes.

The Isua Greenstone Belt contains one of the oldest rock formations on Earth, with study conducted at the rocks revealing the belt to be one of the oldest and best preserved ancient plate tectonic sequences. Plate tectonics is a prominent theory that explains the structure of the Earth's crust and related phenomena, which possibly all began around the Hadean Era, shaping the future of how plates governing landmasses interact to this day. The belt's rock composition includes tonalite, mafic rocks, metasedimentary rocks, banded iron formation, and granodiorite.

Reported in the journal *Nature*, the discovery of these ripples supports theories that life on Earth originated during the so-called Hadean eon. This period in Earth's history, an informal division of Precambrian time, occurred between around 4.6 billion and four billion years ago. It was a time characterized by Earth's initial formation from the accretion of dust and gases, plus frequent collisions of larger planetesimals and other cosmic debris. The period also saw the stabilization of the Earth's core, the development of the crust, and the formation of the atmosphere and oceans.

During this period, impacts from various forms and sizes of cosmic debris released enormous amounts of heat that likely prevented much of the rock from solidifying on the surface. Convection currents in the Earth's mantle brought molten rock to the surface and caused cooling rock to descend into magmatic seas. Heavier elements, including iron, descended to become the core, whereas lighter elements such as silicon rose and became incorporated into the Earth's growing crust. Some scientists believe that the existence of a few grains of zircon, a silicate mineral, which have been dated back to 4.39 billion years ago, confirm the presence of stable continents during that time, along with liquid water and a surface temperature probably less than 100 °C (212 °F).

These zircon grains were found in the Jack Hills, a range of hills in Western Australia, with the true age of the zircon subsequently estimated at 4.375 billion years, with a plus or minus 6-million-year margin of error. While zircons are not technically rocks themselves, the trace elements found within the zircons suggest that they came from water-rich, granite-like rocks, such as granodiorite and tonalite. The discovery is at odds with other findings that suggest that the Earth was initially inhospitable, incapable of supporting the findings that the zircon revealed.

In the Northwest Territories of Canada lies the Acasta Gneiss, a rock formation dated at around 3.58 to 4.031 billion years ago. The formation is comprised of tonalite gneiss, which is mostly composed of quartz and feldspar, and was formed during the Hadean eon. Rock specimens from that period were first unearthed in 1989.

On the coastline of Hudson Bay, Quebec, Canada, a bed of rock exists that dates back 4.28 billion years. In parts of this rock formation, known as the Nuvvuagittuq Greenstone Belt, geologists were able to date the rock using ancient volcanic deposits known as faux amphibolite. After the discovery in 2001, the age of the rocks was contested with different research that dated the rock at between 3.7 billion to 4.37 billion years ago. In March 2017, a report provided evidence that fossils of microorganisms have been found in the Nuvvuagittuq rocks.

Dating from 4.091 billion years ago and found in Antarctica in 1984, Alan Hills 84001 (commonly abbreviated as ALH84001) was a rock that revealed an incredible past. Composed of orthopyroxene, chromite, maskelynite, and iron-rich carbonate, the rock is believed to have originated from Mars. The meteorite gained unprecedented international attention in 1996 when a group of researchers from NASA's Johnson Space Center in Houston announced they had spotted possible signs of Martian life in the meteorite. At the heart of the claim was the discovery of the existence of microscopic magnetite crystals that the researchers claimed resembled ones created by microbes on Earth.

The proposal was widely rejected by the scientific community, with the accepted consensus stating that the original rock did in fact form on Mars around four billion years ago. It was eventually catapulted into space by an impact on the Martian surface that scattered it and many others fragments over a great distance. In the case of Alan Hills 84001, the rock wandered through the Solar System for millions of years before landing on Earth some 13,000 years ago. A number of other meteorites have been discovered and duly certified as heralding from Mars, but Alan Hills 8400 remains by far the oldest, with the next age being just 1.3 billion years ago. The reason for dismissing Alan Hills 8400 as a sign of Martian life involves the debate over the meteorite's unusual features, which were subsequently explained away by methods not requiring life to be present. Still, the meteorite's discovery and the ensuing hype surrounding its confirmed origin on Mars during a period where the red planet contained liquid water remain a quite notable event among the vast array of age-related geological discoveries over the years.

Jumping from Mars to the Moon, we now turn to the anorthosite rock that was brought back by the Apollo 15 mission by astronauts James Benson (Jim) Irwin (1930–1991) and David Randolph Scott, (1932 -), from the Spur crater. The third astronaut on the mission was Alfred Merrill (Al) Worden, who didn't venture down onto the lunar surface, but rather made his contribution to the flight via scientific observations from orbit.

Renamed the Genesis rock after its earlier official stamp of Lunar sample 67,215, there are many reasons why the find was so important, not least of all being that having collected and analyzed the rock, it gave scientists the chance to make the first firm, physical comparison of Earth and Moon rock. When first analyzed, the 4.46-billion-year-old Genesis rock gave indications that it could have originated during a time when the Moon's primordial crust was forming. In a paper published online in the journal *Nature Geoscience*, it was revealed that the Genesis rock and other lunar anorthosites had large traces of water, suggesting that the early Moon was wet when it formed. This went directly against the theory that the Moon was formed from debris generated during a giant impact between the Earth and another planetary body.

Apollo 15 was the ninth manned mission in NASA's Apollo program, and the fourth to land on the Moon. The mission, classed as a J mission (meaning a long stay on the lunar surface), had a greater scientific focus than previous ones. It was a mission that also saw the first outing of the Lunar Roving Vehicle. Launched on July 26, 1971, it was later heralded by NASA as the most successful manned flight ever achieved. Approximately 76 kg of lunar material, including soil, rock, core-tube, and deep-core samples, was returned to Earth.

Life Itself

The various rock formations scattered across the globe tell us a great deal about the Earth's past, some of them perfectly preserving its turbulent history. In many cases, though, the debate is still open, as subsequent pieces of the jigsaw don't always fit so snugly in the overall picture.

In Earth's 4.5-billion-year history, an estimated five billion species have become extinct—that's in excess of 99% of all species that ever lived. And yet, in a report dating from May 2016, an estimated one trillion animals currently exist on our planet. The world is literally teaming with life.

In 2009, after the discovery of a partial skeleton some 16 years previously, "Ardi" was reported to the world's media for the first time, marking a significant moment in our understanding of humankind's evolution.

Around 4.4 million years ago, Ardipithecus ramidus roamed Eastern Africa (Middle awash and Gona, Ethopia) with certain aspects of the skeleton giving a revealing insight into the life of "Ardi" and the possible link that developed into humankind.

The foot bones in her skeleton indicated a divergent large toe combined with a rigid foot. The pelvis, reconstructed from a crushed specimen, was said to show adaptations that combine tree-climbing and bipedal activity. The discoverers argued that Ardi was a human-African ape common ancestor that was not chimpanzee-like. A good sample of canine teeth of this species indicates very little difference in size between males and females.

Ardi's fossils were found alongside faunal remains, indicating that she lived in a wooded environment. This contradicts the open savanna theory for the origin of bipedalism, which states that humans learned to walk upright as climates became drier and environments became more open and grassy.

The team that made the discovery that became Ardi was led by American paleoanthropologist Tim White. The initial finding of a piece of partial hand bone was made by a college student, Yohannes Haile-Selassie.

Over 100 fossil specimens of Ardipithecus ramidus have been uncovered. The discovery name is derived from "ramid," which means "root" in the Afar language of Ethiopia and refers to the closeness of the then-new species to the roots of humanity, while "Ardi" means "ground" or "floor." At the time, the genus Australopithecus afarensis was already established (we will visit this in a moment), so White devised the new genus to distinguish Ardipithecus ramidus from Australopithecus afarensis. Some of the fossil finds predate Ardi—the first ones being found in Ethiopia in 1992—but it was only after 17 years that they were revealed to be of the Ardipithecus ramidus genus.

The female of the Ardipithecus ramidus genus stood at an average of 3 feet 11 inches, weighing in at an average of 110 lbs.

The already established genus Australopithecus afarensis made its mark with the discovery of Lucy. Lucy, the common name for AL 288-1, was comprised of several hundred pieces of bone fossils representing 40% of the skeleton of a female. Discovered in Africa on November 24, 1974, the assembly of bones is also known as "Dinkinesh," which means "You are marvelous." American paleoanthropologist Donald Carl Johnson and graduate student Tom Gray made the discovery along with colleagues, French anthropologist Yves Coppens and French geologist and anthropologist Maurice Taieb. Lucy's name is derived from the Beatles track "Lucy in the Sky with Diamonds," which had been playing loudly and repeatedly after the first day's work on the excavation site. Her bones dated at 3.2 million years old, Lucy is believed to have reached the age of 12, a mature but young adult, although the cause of her premature death remains undetermined. Scientists did however determine that Lucy walked upright and ate what appeared to be a predominately plant-based diet.

Lucy's skeleton is preserved at the National Museum in Ethiopia in Addis Ababa, situated near the Addis Ababa University graduate school. Other exhibits featuring Lucy can be found at the Cleveland Museum of Natural History, the American Museum of Natural History in New York, and the Field Museum in Chicago.

Timeline: Human Evolution

Lucy and others form a link in a chain that many scientists believe forms the evolutionary order from our past to our present. Interpretation and debate remain.

55 million years ago (MYA)—First primitive primates evolve

8–6 MYA—First gorillas evolve. Following the gorillas, chimpanzees and human lineages diverge

5.8 MYA—Orrorin tugenesis. The oldest human ancestor thought to have walked on two legs

5.5 MYA—Ardipithecus (Ardi). An early forest-dwelling human that shares traits with chimpanzees and gorillas

4 MYA—Australopithecines appear. They have brains no larger than that of a chimpanzee but walk upright on two legs. First human ancestors to live on the savannah

3.2 MYA—Australopithecus afarensis. Lucy's time period

2.7 MYA—Paranthropus. The paranthropus dwells in the woods and grasslands. The early human has a massive jaw to chew plants and vegetation. The paranthropus becomes extinct at 1.2 MYA

2.5 MYA—Homo habilis appears. These homonids still retain many ape-like features, but they exhibit a less protruded face. The beginning of stone tool use fashioned from splitting pebbles; the introduction of the Oldowan tradition of tool making that would last a million years. A proportion of hominids develop a taste for meat, which in turn may have contributed to the evolution of larger brains

2 MYA—Homo ergaster. Evidence exists of Homo ergaster in Africa with cranial development suggesting a larger brain capacity

1.8–1.5 MYA—Homo erectus. Found in Asia, Homo erectus has a greater brain size and represents the first true hunter-gatherer ancestor. First predecessor to have migrated out of Africa

1.6 MYA—Potentially first use of fire, notably from the discovery of discolored sediments in Koobi Fora, Kenya, along with charred wood and stone tools found in Israel, dated to 780,000 years ago

600,000 years ago (YA)—Homo heidelbergensis. Lived in South Africa and Europe; had a brain capacity akin to modern humans

500,000 YA—Evidence of purpose-built housing in the form of wooden huts, discovered in Chichibu, Japan. These shelters, discovered on a hillside, predated the discovery of a structure found at Terra Amata in France from around 200,000 to 400,000 years ago. Also discovered on the Chichibu site were 30 stone tools

400,000 YA—Early humans begin hunting their prey with spears

325,000 YA—The oldest surviving human footprints are left by three people who scrambled down the slopes of a volcano in Italy. These early prints, measuring 20 cm long, hint that early humans were less than 1.5 meters tall. Some scientists argue that the oldest human footprints are 3.5 million-year-old tracks found in Tanzania in 1979

280,000 YA—The first complex stone blades and grinding stones emerge

230,000 YA—The first appearance of Neanderthals across Europe, extending from the United Kingdom in the west to Iran in the east. Neanderthals become extinct with the advent of modern humans 28,000 years ago

195,000 YA—Homo sapiens appear. A migration follows across Asia and Europe

170,000 YA—Mitochondrial Eve, the direct ancestor to all living people today. Thought to have lived in Africa

150,000 YA—The development of speech and symbolism. 100,000-year-old shell jewelry discovered

140,000 YA—First evidence of long-distance trade.

110,000 YA—Evidence of the first beads made from ostrich eggshells, carved into a donut-shaped piece, found in Tanzania's Serengeti National Park

50,000 YA—Seen as a landmark time of great advancements. Creation of clothing from animal hides and the development of more highly skilled hunting techniques. Changes in the approach to community and family with the burying of the dead

33,000 YA—Oldest cave art known discovered. Later, Stone Age artisans create the impressive murals at Lascaux and Chauvet in France. Found by three cavers in December 1994, the Chauvet cave drawings are thought to date between 32,000 to 36,000 years old. Contained among the 1,000 of more drawings are depictions of animals, including a mammoths, lions, bears, and even an owl!

18,000 YA—Homo Floresiensis. The so-called "Hobbit" people, found on the Indonesian Island of Flores. Standing at just over one meter tall, these people possess brains similar to the size of chimpanzees, yet despite this, use advanced stone tools

12,000 YA—Modern people reach the Americas

10,000 YA—The development of agriculture becomes widespread with the introduction of villages

5,500 YA—The Bronze Age commences as the Stone Age comes to an end. Humans begin to smelt and work copper and tin. These are formed into implements that eventually replace those made of stone

5.000 YA—Earliest known writing. Possibly dating from a lost Bronze Age Middle Eastern society

4,000 to 3.500 BC—The Sumerians of Mesopotamia develop the world's first civilization.

Chapter 3

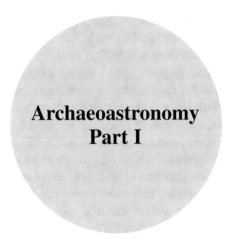

Archaeoastronomy
Part I

It would have been quite a sight, lesson, and honor to observe those who first watched the heavens. For many of them, it was a nightly puzzle that heralded many questions. What are those points of light? Are they gods? Are they good or evil? They move ever so subtly, but do they communicate? Ancient civilizations and their earliest understanding of our Universe had a great impact on how later generations of astronomers across the globe interpreted what was seen in the night sky.

Ancient Egypt

Ancient Egypt, one of the earliest and most advanced civilizations, had a significant and diverse set of religious beliefs that were intertwined with the fabric of society and everyday life, including the patterns and motions of the night sky. A number of Egyptian myths sought to explain and solve the riddles that the heavens above posed. Beyond this sophisticated collection of myths, their society created great temples and the staggeringly wondrous pyramids, all designed and built to with very specific astronomical orientations.

One of the foremost pioneers of archaeoastronomy was British astronomer and scientist Sir Joseph Norman Lockyer (1836–1920). While in Greece on one of his many expeditions (a significant number of which were to observe solar eclipses), Lockyer noticed that the majority of temples in

© Springer Nature Switzerland AG 2018

J. Powell, *Rare Astronomical Sights and Sounds*, The Patrick Moore
Practical Astronomy Series, https://doi.org/10.1007/978-3-319-97701-0_3

the country seemed to have an east–west orientation. Then in Egypt, Lockyer discovered that the temples were oriented to sunrise at midsummer, towards the brightest star in our skies, Sirius, the Dog Star. Lockyer, assuming that the orientation of the Heel-Stone at Stonehenge was also to sunrise at midsummer, calculated the construction of the monument to have taken place in 1680 BC. Some 30 years after Lockyer's death, in 1952, radio carbon dating gave a date for the construction of Stonehenge at 1800 BC.

Many Egyptian buildings were constructed with astronomy in mind. Temples and pyramids alike were placed in relation to the heavens above, including the stars, zodiac, and constellations. In different Egyptian cities, different patterns of orientation and alignment were observed, this being governed by the religion that was beholden to that specific city. Still, many of their underlying principles remained the same; for instance, they were often purposefully aligned with a star that rose when it was necessary to start seeding the land, in preparation for crop growth in the months that followed. Other temples and pyramids were aligned with stars thought to govern the time to harvest the crop. This marking of time with the passage of the night sky acted as an accurate and stable schedule for cultivating the land, a practice that was abided by religiously.

Other buildings were orientated toward the solstices or equinoxes. It is estimated that as early as 4000 BC, temples were built so that only one particular room within their walls would receive sunlight at only one precise time of the year. In some cases, the building method allowed for a gradually narrowing succession of doors, which ultimately led into a specific room in which the one beam of sunlight would eventually concentrate on the image of a god or goddess. These designs were in some cases quite elaborate, a historical accolade to the architects who created them under the governance of the astronomers, and the highly skilled planners and builders who made the vision a reality.

One such complex design is at the temple of Medinet Habu, which stands on the West Bank of Luxor. Here, there are actually two buildings that are slightly off kilter. It has been suggested that the second structure was built when the altitude of the other temple's orientation stars changed over a period of time.

Through further study, Lockyer subsequently broke down ancient astronomy into three distinct phases:

1. A civilization goes through the worship stage, where astronomical phenomena are viewed only as actions, moods, and warnings of the gods.
2. A civilization progresses to using astronomy for terrestrial purposes, such as for agriculture and navigation.
3. A civilization studies astronomy solely for the purpose of gaining knowledge.

The Ancient Egyptians started in the initial worship phase, with the numerous gods and goddesses pictured in many paintings and murals depicting life at the time. Although the Egyptians were to move on to the second phase by involving astronomy in their everyday lives, the idea of associating certain gods with the constellations became engrained in their culture.

For example, the constellation of Orion was represented by Osiris, the god of the afterlife, the underworld, and the dead, and more appropriately the god of transition, resurrection, and regeneration. Osiris was considered to be not only a merciful judge of the dead in the afterlife, but also the underworld agency that granted all life, including the sprouting of vegetation and the fertile flooding of the Nile River. Representing the hope of new life after death, Osiris became more and more closely associated with the cycles of nature, with the Egyptians noting in particular the link between the growth of new vegetation and the flooding the Nile, and the heliacal rising of Orion and Sirius at the start of the New Year.

The Nile was at the very center of the Egyptian civilization, with its annual floods providing rich soil for enough agriculture to support the growing population. Egyptian priest astronomers (the two roles were closely intertwined) were given high stature within the community for their ability to predict exactly when the Nile would flood, a feat that if closely investigated by any layperson would have yielded the revelation that it always occurred around the summer solstice.

The stars in Ancient Egyptian astronomy were represented by Seshat, the goddess of wisdom, knowledge, and writing. Seshat was seen as a scribe and record keeper, her name literally translating as "she who scrivens," or "she who holds the scribe."

Beyond her role in the astronomy domain, Seshat was identified as the goddess of accounting, building, mathematics, and surveying. A pattern seen on one of her depicted garments is that of a spotted feline. The pattern, on natural hide, was thought to represent the stars, being a symbol of eternity and carrying a strong association with the night sky. Other gods and goddesses were represented by actual astronomical bodies, the Moon according to some being Thoth, God of magic and writing.

The horizon held particular significance for the Egyptians, for it was here that the Sun appeared and disappeared. The Sun represented light, warmth, and growth, being seen as the ruler of all that it created. Any deity associated with the Sun was thus very important.

Numerous gods were attached to the Sun, depending on its position in the sky. At dawn and during the morning, the Sun was Horus, one of the most significant of Ancient Egyptian deities. Horus was the son of Isis and Osiris.

The noon Sun was represented by Ra, with the disk of the Sun itself seen as the body or eye of Ra. The evening Sun became the god Atum, the god responsible for lifting pharaohs from the tombs to the stars above. The color red, which is sometimes seen at sunset, was considered by the Ancient Egyptians to be the blood from the Sun god as he died. To complete the cycle, after the Sun had set, the Sun itself then became Osiris, god of death and rebirth. Hence, night was associated with death, and the day with life or rebirth, making for a continual cycle that reflected the Egyptians' idea of immortality.

Ancient India

Dating back almost 4,000 years, the practices of astronomy and astrology in ancient India stem largely from the Sanskrit sacred books known as the Vedas. However, there are also references in history that trace the earliest roots in Indian astronomy to the period of the Indus Valley Civilization, or even earlier. The Indus Valley Civilization (Bronze Age), or Harappan Civilization, dates from 3300–1300 BCE. Along with Ancient Egypt and Mesopotamia, it is one of the three early cradles of civilization of the Old World. At its peak, the Indus Valley Civilization may well have seen its population exceed five million.

Astronomy later developed as a discipline of Vedanga or one of the auxiliary disciplines associated with the study of Vedas. The oldest of these is known as the Vedanga Jyotisha, dated to 1400–1200 BCE. Jyotisha, one of the six ancient Vedangas connected with the Vedas, is the science of tracking and predicting the movements of astronomical bodies in order to keep time. The timekeeping was used to fix the days and hours of Vedic rituals. The text within the Jyotisha discusses how the movement of the planets, Sun, and Moon form a calendar. These movements are calculated with the aid of trigonometry and mathematical formulae.

The religious texts known as the Vedas were a series of hymns composed over a period of hundreds of years. Within the writings, the reader is offered an insight into how the night sky was perceived by its authors. As in most of the ancient world and its tightly interwoven culture, events in the sky were thought to have a direct effect on the people. Astronomy as a science was born from those who studied the motion of the celestial bodies.

The Vedas are thought to be associated with a large group of nomads called the Aryans who, traveling from central Asia, embarked on a trek that was to take them across the Hindu Kush Mountains, migrating their way into the Indian subcontinent. Considering the Hindu Kush Mountains extend 800 km west from Pamir Knot in northern Pakistan into northeast

Afghanistan, the journey must have been daunting. With the highest peak measuring 7,692 meters in Tirich Mir, these permanently snow-covered mountains would have offered little vegetation or comfort to the traveler.

The sheer number of Aryans making the ascent into the Indian subcontinent (the exact amount remains undetermined) is viewed by some as an invasion, for it was not long that that the Aryan language gained supremacy over other regional dialects. This would indicate an influx of a population from perhaps an entire nation state into the foreign land.

The Veda texts call the gods Devas, which means "bright" and refers to the luminous nature of the Sun and stars. The Sun itself, along with comets, the sky, dawn, and the horizon were all deified based on their aspects. Not a great deal is known or included about the authors of the Vedas themselves. Perhaps this was done to avoid readers focusing too heavily on the messenger who is delivering the knowledge. The overview of the text is predominately spiritual, reflecting the worldview of the period. Despite the dominance exerted by the Aryans upon the Vedas, elements of the natives of India before the Aryans remain evident. These elements come in the guise of references to animistic and totemic worship of many spirits, with these spirits dwelling in stones, animals, trees, rivers, mountains, and the stars.

The Vedas, or Veda, meaning "knowledge," were composed in Vedic Sanskrit and constitute the oldest layer of Sanskrit literature, and in the study of Hinduism, the oldest known scriptures. They were first dispersed in oral form before actually being written, incorporating mythological accounts, poems, prayers, and formulas within the text. The passing of this "knowledge" originated in the northwestern region of the Indian subcontinent (present day Pakistan). It was from here that the teachings spread, passing down from generation to generation before eventually being committed to writing.

The basic Vedic texts are the Samhita "Collections" of four Vedas:

- **Rig-Veda:** "Knowledge of the Hymns of Praise"
- **Sama-Veda**: "knowledge of the Melodies," used for chanting
- **Yajur-Veda:** "Knowledge of the Sacrificial Formulas," used for liturgy
- **Atharva-Veda:** "Knowledge of the Magic Formulas," named after a kind of group of priests

Rig-Veda

The Vedas were composed at different periods. The oldest is the Rig-Veda, and it is here that we find the early mentioning of astronomy. Written in a very obscure style and filled with metaphors and allusions, the Rig-Veda considers the Earth to be in a shell supported by elephants. The elephants were a symbol of strength and were themselves supported by a tortoise, representing infinite slowness.

The Rig-Veda is considered to be the most important text of the Vedic collection, not the least because of its size, comprising 1,028 hymns and divided into 10 books called mandalas. The Rig-Veda deals with many gods, but there are a few who govern a substantial number of the 1,028 hymns. The Rig-Veda contains many significant hymns, such as the Purusha sukta, which gives a description of the spiritual unity of the universe, and the Nasadiya sukta, which is also known as the "Hymn of Creation."

The three most acknowledged gods within the Rig-Veda include: Indra, the storm god, who the Vedas describes as the god "Who wields the thunderbolts." Agni, the god of fire, commands less material (200 hymns), but despite this lower count, he is often referred to in Vedic literature as the most important god. This is because he is associated with the flame that ultimately lifts any sacrifice to heaven, denoting a fiery life and spirit of the world, the "vital spark," the principle of life in animate and inanimate nature. Agni was viewed as a kind of messenger, carrying souls from the realm of the living to the realm of the dead. Cremation was believed to prevent the spirit of the dead remaining among the living, which is why those who worshipped Agni burned their dead. The third god, Soma, is the personification of the sacred soma plant, the juice of which was thought to be holy and intoxicating to gods and men. With just over 100 hymns, Soma was the god of sacrifices and was in some cases associated with the Moon. Soma was a fermented juice drink, which was believed to have been consumed by the Hindu gods and their ancient priests during rituals.

This trio of gods holds the highest esteem within the Rig-Veda in terms of the sheer number of hymns dedicated to each one.

Sama-Veda

The Sama-Veda has verses that are almost entirely raised from the Rig-Veda. However, for the purpose of being used in chant form, they are somewhat rearranged so as to make for a more melodic tone.

Yajur-Veda

The Yajur-Veda is divided up into the White Yajur-Veda and the Black Yajur-Veda, with prose commentaries on how to enact religious rituals and sacrifices. According to one source, this Veda is said to be similar in function to the ancient Egyptian Book of the Dead.

Atharva-Veda

The Atharva-Veda contains charms and magical incantations, and generally has more of a folkloristic style to its text. The youngest of the four, the Atharva-Veda is considered by some to not have any connection with the Vedas at all. Nevertheless, in the study of Vedic history and sociology, it is considered by many to be next in importance to the Rig-Veda.

The Progression of Indian Astronomy

As time progressed, Indian astronomy began to evolve out of its spiritual grounding, taking on more and more of a scientific guise. Early astronomers began to make clear distinctions about certain phenomena, most notably at the beginning of the first century. As astronomy began to branch out, moving away from the Vedas, it entered a period known as the Siddhantic era. The era commenced with three books called the Siddhanat, or "Solutions."

It is in these "Solutions" that astronomers begin to recognize that the stars are the same as our Sun, only more distant in the heavens. Particular texts from the time note the night sky as being full of suns, and that when our own Sun goes below the horizon, a thousand suns take its place. This finding may not be considered in modern times to be a massive leap in understanding, but during this dawning of astronomy in Indian scientific groups, it was an important revelation that provided a firm foothold on the ladder to more far-reaching discoveries. The notion that the Sun was at the center of the universe was still commonly held, being challenged only by a select few Greeks at the time. Some may conclude from this that the Greeks were more advanced in their understanding of astronomy, but in fact, most Greeks still used the concept of celestial crystal spheres to explain the cosmos. First century Indian astronomers, having already established that the Sun was one of many stars, understood that the Earth was spherical and even attempted to calculate the circumference of the planet.

Early Siddhantic era texts and accounts are poorly documented, although there is one intriguing reference by Yajnavalkya, one of the foremost figures in the earliest period of Indian astronomy, estimated to have lived in the 8th or 7th century BCE. Yajnavalkya, who is considered one of the earliest documented philosophers in history, discussed the concept of a heliocentric universe. Vedic literature also connects him with references to the motion of the sun. Yajnavalkya proposed that a cycle of 95 years was required to reconcile the lunar and solar years, indicating that the length of the year was known to a great degree of accuracy. Yajnavalkya also first proposed that the circuit of the Sun was asymmetric in its four quarters. The proportion for the two halves of the year described by him was 176:189. This proportion is most interesting, as it also describes the exact same asymmetry of the two halves in the Angkor Wat temple in Cambodia.

The Angkor Wat temple (also known as "Capital," or alternatively, "Temple City"), is the largest religious monument in the world and was built by the Khmer King Suryavarman II in the early 12th century in Yasodharapura, modern day Angkor. The temple was unique at the time, breaking from the Shaivism tradition of previous kings by being dedicated to the supreme solar god, Lord Vishnu.

Encompassing an area of 200 hectares, the temple has a 65-meter-high central tower surrounded by four smaller towers and a series of enclosure walls. The temple itself is surrounded by a 200-meter moat that encircles a perimeter of more than five km. The moat is four meters deep and during its construction would have helped stabilize the temple's foundations, preventing groundwater from rising too high or falling too low.

The Khmers adhered to the Indian belief that the temple must be constructed and built in alignment to a mathematical system so that it could function in harmony with the universe. This mathematical system was indeed borne out, as certain distances between architectural elements within the temple reflect numbers that relate to Indian mythology and cosmology.

This staggering paralleling within the temple of its architecture and infusion with astronomy is considerable, with the solar axis of the temple leading directly to the central sanctuary, a sanctum sanctorum devoted to Vishnu. The striking balance continues with lunar alignments along the western and eastern axis of the temple. Archaeologists discovered that when added together, the dimensions of the highest rectangular level of the temple equal 365, the number of days in a year. Additionally, in the central sanctuary, the distances between sets of steps count out the lunar cycles, or synodic months. Also, the lengths and widths of the central tower add up to approximately 91. On average, there are 91 days between any solstice and the next equinox, or any equinox and the next solstice.

Given these fascinating insights, the general understanding is that the temple is in fact an ancient observatory, designed, planned, and then constructed to link many heavenly aspects to everything that lies within it.

The theory of a heliocentric universe was revived in the fifth century CE, when astronomical records first became properly recorded by Indian astronomer and mathematician Aryabhata (b.476 CE). Aryabhata advanced the heliocentric theory by employing a more rigorous mathematical approach to astronomy. His methodology also led to the theory that the Sun is the source of moonlight, and that the Earth itself rotates, rather than the skies.

Additionally, Aryabhata formulated ways to forecast eclipses using mathematical models. His ideas and theories are found in his book, the *Aryabhata*, which went on to influence European mathematicians and astronomers when it was translated into Latin in the 13th century.

Long before Sir Isaac Newton (1643–1727), Indian astronomers in the sixth century, and in particular Varahamihira (b.476 CE), came up with the idea that the same force holding objects to the Earth was also responsible for holding celestial bodies in place. Varahamihira concluded that there must be some type of attractive force keeping such objects stationary. Varahamihira is responsible for the compilation of a compendium of Greek, Egyptian, Roman, and Indian astronomy, with the text incorporating complex mathematical charts and tables.

In the seventh century, astronomer Brahmagupta attempted to calculate the circumference of the planet. Brahmagupta arrived at the figure of 36,000 km—very close to the actual figure. Brahmagupta made important contributions on many levels, with his methods for calculating the position of heavenly bodies over time setting a standard for determining conjunctions and both solar and lunar eclipses. Brahmagupta also challenged the Vedas over the idea that the Moon was farther away from the Earth than the Sun. He explained that because the Moon is in fact closer to the Earth than the Sun, the degree of the illuminated part of the Moon depends on the relative positions of the Sun and the Moon, and this can be computed from the size of the angle between the two bodies. Another accolade of note is that Brahmagupta was the first to create rules to compute with the numeral zero!

Indian astronomy continued to flourish as the science entered the post-Siddhantic era, brought about by the rise of Islam in India and the introduction of Arabic-translated Greek texts into the science community. For this reason, the post-Siddhantic era is often also known as the Zij era, after the Zij tablets containing astronomical data. Al-Khwarizmi (780–850 CE) was responsible for translating the early Zij tablets from Sanskrit. Beyond notable contributions to mathematics, he revised Ptolemy's *Geography* and wrote on both astronomy and astrology.

The ongoing progress of the Indian astronomers helped form the cornerstones of Islamic astronomy. The great Mughal Era of India (1526–1725) saw the union of Hindu mathematical techniques with Islamic observational techniques, prompting great advances in astronomy. Mughal astronomers did not appear to have much concern for the theoretical side of astronomy, with true progress shown on an observational level. Through the British colonial annexation of India, these combined techniques became a major part of modern astronomy.

Mesopotamia

Astronomy in Sumer, Babylon, and Assyria began in the early settlements of the agricultural societies as such civilizations grew and spread across the lands over time. Mesopotamia, the land between the River Tigris and Euphrates, was the birthplace of civilization almost 10,000 years ago, and it is in Sumer that we find the oldest records of the study of astronomy.

The Sumerian region was the first to establish itself in astronomy, with Babylonians and Assyrians inheriting its wealth of knowledge as they too established themselves as civilizations in the same geographical area over the years that followed. Along with the scientific data collected by the Sumerians through observational work, the Babylonians and Assyrians acquired many myths and legends, which, like those of the Sumerians, became embedded in their society's culture and belief system. These civilizations intertwined fact and myth in ways that were eventually passed down into Greek culture and remain present even today. The Sumerians and Babylonians should be given most of the credit, for they gave us many of the constellation names that are still in use in modern day astronomy, through their carefully logged accounts of night sky activity—a very precious record from that era.

Between 2000–3000 BCE, the Sumerians and Babylonians named constellations such as Leo, Taurus, Scorpius, Auriga, Gemini, Capricorn, and Sagittarius. The zodiac was created by the Babylonians, with 12 constellations all individually marked out in the night sky. Well-preserved records include references made to the constellations that travel the zodiac, with accurate and precise recordings of all movements. The very movement of the constellations from season to season proved to have many practical applications. This month by month transition gave the farming community an annual, reliable idea of when to plant and harvest crops, a familiar pattern across so many ancient cultures. The Babylonian accounts are quite remarkable, with written records existing of calendars used for the planting of crops.

As with many cultures, the basis for astronomy was initially formed within the realm of astrology. In the ancient world of Mesopotamia, occurrences and strange happenings in the sky were linked with events on Earth, often having profound religious meanings for the population. Indeed, so powerful was the connection that the astronomical events affected far more than simply religion, with social and political aspects being affected as well.

The monarchy and royal subjects relied heavily on omens, with the power of reading such omens bestowed on a group of so-called priest-astronomers. The priest-astronomers' duty to interpret astronomical events gave them an almost unchallenged status within society, for they dictated how people should live, how they should act, and what they should do, all based on how the priest-astronomer reads meaning in the night sky.

All of what we know about the ancient world of Mesopotamia relies almost entirely records kept throughout the period. Without them, much of what we understand would have been left to the interpretation of others, with the original sources becoming distorted, diluted, and perhaps manipulated in some way. Take the great fire at the Ancient Library of Alexandria, which destroyed so many promising texts that we now only know exist from secondhand sources.

Records for the Mesopotamian era were mostly kept on clay tablets through a writing system known as cuneiform. The tablets bear distinctive wedge-shaped marks, the impressions having been made by a blunt reed that would have been used like a pen or stylus, creating word-signs (pictographs). Cuneiform literally means "wedge shaped," from the Latin word "cuneus," and it is believed that the Sumerians around 3500–3000 BCE were responsible for first introducing it as means of making records. The introduction of cuneiform by the Sumerians is deemed one of its most significant contributions to civilization. The history of this type of record can be traced back to the tablets that preceded the cuneiform, the marks of which are known as proto-cuneiform. While still incorporating pictographs, the subject matter and format was less intricate, as only less subtle shapes were addressed, representing things like a king or a great flood. This method later developed into a style that was able to produce a much wider and broader depiction of events, with a word-concept allowing the writer to express more nuanced meanings, and giving the reader a more tangible and individualist feel for what was being written. The development continued with characters able to define much more clearly why they were being used and in what context.

This particular advancement is evident in the works of Enheduanna (2285–2250 BCE), who wrote a series of hymns in cuneiform. The hymns were so sophisticated that they were able to convey emotional states of mind such as love and adoration, betrayal and fear, longing and hope, with the writer even able to express accurate and precise reasons for the inclusion of such emotions in the text. Enheduanna, who is believed to be the world's first author known by name, is credited with creating the paradigms of poetry, psalms, and prayers that were used throughout the ancient world, which, over the centuries that followed, led to the development of the genres recognized in the present day.

Enheduanna's influence is staggering, with her works altering the established nature of the Mesopotamian gods and the perception of the divine within the community. Her name, which translates as "High Priestess of Anu" (the sky god), became legend. Aside from her hymns, she is remembered for 42 poems she wrote reflecting her own personal frustrations and hopes. These poems illustrate her religious devotion, her response to war, and her own feelings about the world in which she lived, giving precious insights beyond a mere factual account of the time.

The Babylonian records kept on cuneiform were initially used for business purposes, such as to keep track of any financial transactions or to keep accurate inventories. However, some tablets have been discovered that have scientific topics within their text. One notable example of this is the Venus Tablet of King Ammizaduga, which contains the scientific methodology used by Babylonian astronomers in the study of the planet Venus. The tablet records the rise times of Venus and its first and last visibility on the horizon before or after sunrise and sunset in the form of lunar dates. Another astronomical cuneiform tablet was found in the tomb of King Ashurbanipal of Nineveh, who ruled the Neo-Assyrian Empire from 668–627 BCE. This tablet also details the times of Venus as a morning and evening apparition.

The Babylonians were able to recognize that Venus, whether observed in the evening or in the morning, was in fact the same object, and while this may sound quite basic in understanding the workings of our Solar System, at the time, it was quite profound. The Babylonians also developed a method for calculating the length of the Venus cycle. Astronomers established that the length of one cycle was 587 days, compared with the actual value of 584 days— quite a remarkable achievement. The Babylonians only slightly miscalculated, attempting to coincide the cycles of Venus with the phases of the Moon. Several cuneiform tablets show that the Babylonians and the later civilization of the Assyrians were able to predict lunar eclipses by engaging a simple method that calculated future predictions based upon past observations. The tablets list a series of lunar eclipses, with a marking of the time between successive events clearly recorded.

Beyond these innovations in calculating astronomical events, one of the most significant breakthroughs for modern science was the invention of a degree system, which allowed astronomers to distinguish with accuracy the positions of celestial bodies in the sky. The system, which was later adopted by the Greeks, engaged a method similar to our present-day use of degrees to calculate latitude and longitude.

Arab and Islamic Early Astronomy

During a period in history when Western civilization was experiencing the Dark Ages, from roughly the fifth to tenth centuries, an Islamic empire stretched from Central Asia to Southern Europe. While much of Europe was undergoing a suppression of intellectual thought and conflicts over religion, something else was taking place in the Islamic Empire.

As the Dark Ages descending on the Europe, a "Golden Age" emerged in the Islamic world, with astronomy of particular interest to scholars in Iran and Iraq. Both mathematics and astronomy rapidly developed at this time, with the whole involvement of the scientific community encouraged as a new era of learning and development in the sciences sprung forth.

At the time of this Golden Age, one of the few astronomical textbooks that had been translated into Arabic was Ptolemy's *Almagest*. Claudius Ptolemy (100–168 CE) was a Greco-Roman mathematician and astronomer who lived in Alexandria in the Roman province of Egypt. The *Almagest* is the only surviving comprehensive ancient treatise on astronomy. Ptolemy's model of an Earth-centered universe formed the basis of Arab and Islamic astronomy. Over time, several Islamic astronomers went on to make observations and calculations relating to the night sky that were far more accurate than Ptolemy's own.

Not all astronomers having read and understood Ptolemy's *Almagest* actually agreed with. Questions and criticisms came from a number of quarters, such as Abu'l-Hassan Ali ibn Abd al-Rahman ibn Ahmad ibn Yunus al-Sadafi (950–1009 CE) from Egypt, who found faults with Ptolemy's observations about the movement of the planets and their eccentricities.

Ptolemy had attempted to find an explanation for how the planets orbited in the sky, and why at certain times they did not behave as expected, like how some seemed to be ordered in their orbit and others displayed a more elaborate nature in their movements. Ptolemy also tried to explain how the Earth moved within all of these parameters when associated with the other planets of the Solar System. Ptolemy calculated that the wobble of the Earth—precession as we know it—varied by one degree every 100 years. Ibn Yunus studied the observations and calculations made by Ptolemy at length, pronouncing a glaring inaccuracy and stating that it was in fact one degree in 70 years. However, neither Ptolemy nor Ibn Yunus during their time were able to attribute this change to the Earth's wobble, because during both their time periods, it was still believed that the Earth was at the center of the universe.

Ibn Yunus went on to write many works on astronomy, his most notable contribution being a handbook of astronomical tables that contained very accurate observations, many of which Ibn Yunus had obtained from very large astronomical instruments. It is thought that Ibn Yunus used a pendulum for time measurement, although there appears to be some division over whether or not he really did so. In total, Ibn Yunus described 40 planetary conjunctions and 30 lunar eclipses.

Another voice heard in the Islamic community during this time was Ala Al-Din Abu'l Hasan Ali ibn Ibrahim ibn al-Shatir (1304–1375 CE), whose most notable contribution to astronomy was his treatise *kitab nihayat al-sul tashih al-usul* (the Final Conquest Concerning the Rectification of Principles). The work immeasurably transformed the Ptolemaic models relating to the Sun, Moon, and planets. Born in Damascus, Ibn al-Shatir was responsible for the regulation of the astronomically defined times of prayer, drafting a set of tables with the values of certain spherical astronomical functions that directly affected the prayer times.

Collectively, these and other astronomers and mathematicians altered the field of astronomy forever. Islamic astronomy was built on the sciences of two great cultures, the Greeks and the Indians, and it in turn made a dramatic and powerful influence on the cultivation and expansion of ideas that would lead the world into the beginning of the Renaissance period.

The Islamic religion requires believers in the faith to pray five times a day at specified positions of the Sun. The most accurate way to do this was to use astronomical time-keeping, offering precision and pinpoint efficiency so that prayers could be observed with mastery. System was also used to identify religious festivals. The Koran makes frequent references to astronomical patterns visible in the sky and presents itself as a major source of the traditions associated with Islamic astronomy.

Another important religious use for astronomy in Islam was the determination of longitude and latitude. Using the stars, and in particular the Pole Star, as guides, several tables were compiled that calculated the longitude and latitude of important cities in the Islamic world. Using the tables as reference, Muslims could therefore be assured that when prayer time was upon them, they could always be praying in the direction of Mecca, as specified in the Koran.

Beyond religious uses, astronomy was utilized as a navigational tool. One such instrument, devised by the Greeks and perfected by the Arabs, was the astrolabe. The astrolabe, invented around 200 BCE, is often credited as the work of Greek astronomer, geographer, and mathematician Hipparchus (190–120 BCE). The astrolabe calculated positions of certain stars in order to determine direction. In essence, it can be described as a very ancient astronomical computer, given its ability to solve problems relating to time and the position of the Sun and stars. After being perfected in the Islamic world, the astrolabe was eventually introduced into Europe in the early 12th century. It became the most popular, preferred instrument until around 1650, when even more specialized instruments superseded it. The astrolabe's contribution to the astronomical world was invaluable.

The pursuit of astronomy and other sciences was given great support from nobility of the time, with rewards bestowed on scientists for achievements produced while serving in the courts held by individual leaders across the land. In 830, the Khalifah al-Ma'muun founded Bayt-al-Hikman, the "House of Wisdom," as a central gathering place for scholars across the empire. This learned group combined its knowledge to translate many Greek and Persian texts into Arabic so that the findings could be discussed, possibly challenged, and elaborated upon. The building, which was initially constructed as a library, became an unrivaled center for the study of humanities and for science in medieval Islam, covering such topics as mathematics, medicine, alchemy and chemistry, zoology, geography, and cartography. By the middle part of the ninth century, the House of Wisdom had the largest selection of books in the world. Sadly, this collection was destroyed following the Mongol Siege of Baghdad in 1258. This siege is considered to have marked the end of the Golden Age.

The Golden Age had delivered a truly defining era in the Islamic study of astronomy, with many great names in the field leaving an imprint on history throughout the period. One such Islamic astronomer was Abu Ja'far Muhammad ibn Musa al-Khwarizmi (780–850 CE). Al-Khwarizmi invented algebra completely in words rather than mathematical expressions, basing the system on the Indian numbers borrowed by the Arabs, which we term today as Arabic numerals. His contributions were far reaching, including his construction of a table of latitudes and longitudes that encompassed 2,402 cities and landmarks, forming the basis of an early world map. His work was translated into Latin hundreds of years later and served as an introduction to the world of the Indian number system, including al-Khwarizmi's concept of zero. Amongst his other studies were detailed calculations of the positions of the Sun, Moon, and planets, with a number of eclipse calculations. Al-Khwarizmi also revised Ptolemy's *Geography*, assisting in a project to determine the circumference of the Earth and overseeing the work of 70 geographers.

Another great Islamic astronomer to rise to prominence during this era was Abu'l-Abbas Ahmad ibn Muhammad ibn Kathir al-Farghani (800–870 CE). Also known as Alfraganus in the Western world, he was responsible for writing extensively on the motion of celestial bodies. Around 833, his textbook *Kitab fi Jawami' 'Ilm al-Nujum* ("Compendium of the Science of the Stars" or "Elements of Astronomy on Celestial Motions") was a competent descriptive summary of Ptolemy's *Almagest*. Like other Islamic astronomers, he accepted the Ptolemaic model of the universe, but he went further, incorporating the findings and calculations of his colleagues into his textbook, making for a revised and updated version of Ptolemy's original work.

Ibn Kathir al-Farghani's work was translated into Latin in the 12th century, helping to spread Ptolemaic astronomy throughout Europe. The work had a substantial impact on a European world thirsty for knowledge. It is said that Italian poet Dante Alighieri (1265–1321) sourced much of his astronomical knowledge for *The Divine Comedy* from the text of ibn Kathir al-Farghani. Later christened *Divina* by Italian writer and poet Giovanni Boccaccio (1313–1375), Dante's *Divine Comedy* is widely considered to be the most important poem of the Middle Ages. In the 15th century, Christopher Columbus (1451–1506) used ibn Kahtir al-Farghani's estimates for the Earth's circumference as the basis for his voyages to America. However, Columbus made the mistake of taking ibn Kahtir al-Farghani's 7,091-foot Arabic mile to be a 4,856-foot Roman mile—an error that caused him to underestimate the Earth's circumference and made him believe he could take a shortcut to Asia. Given the success of his travels, this turned out to be but a mere drop in the ocean of Columbus' collective achievements.

Abu Mahmud Hamid ibn Khidr Khojandi (900–1000 CE), known also as Abu Mahmood Khojandi or al-Khujandi, was a Central Asian astronomer and mathematician who lived in the 10th century. While few facts are known about his life, we can discern from some of his works a little about the man himself. His contribution to astronomy was significant, marked with his building of a huge observatory near Tehran, Iran. He is known to have constructed a huge mural sextant in 994 CE, intended to determine the Earth's axial tilt. He determined the axial tilt to be 23 degrees 32' 19" in the year 994, a staggering feat, considering that the precise axial tilt known to us in modern times is 23 degrees 34'. Al-Khujandi noted that compared to his own calculations, earlier astronomers had found higher values. In analyzing the data, he discovered that the axial tilt is not constant but is in fact (currently) decreasing. The two-second variance in al-Khujandi's calculations and the modern-day known axial tilt was probably due to the weight of the instrument settling over the course of the observations. Al-Khujandi, was also the first astronomer capable of measuring to an accuracy of arcseconds, and he compiled a list of latitudes and longitudes of major cities.

Aside from compiling many astronomical tables, Persian scientist, philosopher, and poet Omar Khayyam (Ghiyath al-Din Abu'l-Fath Umar ibn Ibrahim al-Nisaburi al-Khayyami) (1048–1131) made a remarkable reformation of the calendar. His calculations were more accurate than the Julian calendar and came close to the Georgian calendar. His astonishing work proved to be a very precise computation of time, calculating that a year was 365.24219858156 days long, which is accurate to the 6th decimal place. Known as the Jalali calendar after the Sultan Malik-Shah, who commissioned al-Khayyami to build an observatory and reform the Persian calendar, it entered the arena alongside other proposed calendars that had been produced by fellow mathematicians and astronomers at the time, all of which eventually had to bow their heads to the exactness of al-Khayyami's work. Apart from being one of the oldest calendars in the world, it remains one of the most accurate solar calendars still in use.

The Jalali calendar was a true solar calendar, where the duration of each month is equal to the time of the passages of the Sun across the corresponding signs of the zodiac. The calendar introduced a 33-year intercalation cycle, which consisted of 25 ordinary years measuring 365 days each, and eight leap years measuring 366 days. With a panel of eight scholars working under al-Khayyami, the group implemented an intercalation system on the calendar based on quadrennial and quinquennial leap years.

Ancient Chinese Astronomy

Unlike cultures that focused on mainly the religious aspects of the sky, early Chinese civilizations set about using astronomy for practical purposes from the very beginning. This is not to say that astrology did not play a part in such studies, but that from early on, there seemed to be a distinct distancing between the work of astronomers and astrologers. As astronomers began to chart regular events like lunar eclipses, astrologers found such happenings losing importance as potential omens, thereby weakening their own power under the Emperor.

Ancient Chinese astrologers developed what was known as the "yellow path," the Chinese version of the zodiac introduced to help guide the life of the people on Earth. It references the Sun traveling along the ecliptic, entering the 12 houses along the yellow path, each of which was mirrored in Western astrology. The Chinese followed a calendar of 12 lunar months and calculated the year to be 365.25 days long. They translated this "magic" number into a unit of degrees by setting the number of degrees in a circle equal to 365.25 (as compared to 360 degrees). The Chinese also divided the sky up into four quarters, with seven mansions in each, making 28 in total. These were used to chart the position of the Moon as it crossed the sky.

The first Chinese records of astronomy date from around 3000 BCE. The first human record of an eclipse was made in 2136 BCE. Over hundreds of years of advanced sky-watching and detailed record keeping, the Chinese became very adept at predicting lunar eclipses. They used circumpolar stars as their reference point for the heavens, unlike the Indo-Europeans, who based their calculations upon the rising and setting of celestial bodies on the ecliptic and the horizon. The circumpolar constellation of the Big Dipper helped mark the passage of time and the seasons, using the position and orientation of the formation relative to the Pole Star during any given evening.

While detailed studies of the sky and making records of observational work formed the main body of the astronomer's work, it was not at the core of the astronomer's duty. Instead, the Chinese obsession of accurately charting time was paramount. Astronomers had to announce the first day of every new month and predict lunar eclipses. Meticulous studies of the sky allowed them to do this with more than a fair degree of accuracy. The failure to do so often resulted in dire consequences for the astronomer, in most instances ending with a beheading.

To make their calculations, Chinese astronomers developed a process whereby the sky was equally divided into 12 branches and 10 stems all arranged around the ecliptic, creating a 60-year cycle. The credit for the implementation of this system lies with Emperor Huang Ti, "Huang" meaning "Yellow" (he was known as the Yellow Emperor), whose reign began in about 2607 BCE. It is thought that he was also responsible for the building of a great observatory and planetarium that helped make observations more accurate. Huang Ti's reign is a legendary one, credited with the introduction of wooden houses, carts, boats, and the bow and arrow. Writing and the use of coined money was also accredited to the Emperor's reign. His wife too was said to have discovered sericulture, teaching women of the era how to breed silkworms and weave silk fabrics.

The accuracy of Chinese recordings has become legendary throughout the world, with ancient Chinese astronomers making detailed studies of such phenomena as comets, sunspots, solar flares, and novae— all long before any other culture had done so. There was even an attempt to catalogue every single star in the night sky, defining the constellations by one governing bright star, which astronomers referred to as the king, and all the fainter associated stars within the grouping known as princes.

One such master of cataloguing the skies was astronomer Shi Shen, who is credited with the earliest intentional sunspot observation. Shen also accurately noted the positioning of 121 stars. His work is recorded in texts such the eight-volume *Astronomy*, the one-volume *Celestial Map*, and the one-volume *Star Catalogue of Shi*. The latter two are believed to have been written by a school of Shi Shen followers. These texts are critical to our understanding of how the Chinese arrived at such observations with so much detail. Sadly, much of Shen's work did not survive the passage of time.

To have made such observations during this time, Ancient Chinese astronomers must have been equipped with instruments such as the armillary sphere. The sphere, whose variations include the astrolabe, is a model of objects in the sky on the celestial sphere. It consisted of a spherical framework of rings, centered on the Earth or the Sun, which represent lines of celestial longitude and latitude and other astronomically important features, such as the ecliptic. This model differs from a celestial globe, which is a smooth sphere whose principal purpose is to map the constellations. The armillary sphere is apparently equipped to do much more that the celestial globe. With the Earth at the center, an armillary sphere is known as *Ptolemaic*; with the Sun at the center, it is known as *Copernican.*

One of the most famous observations made by Chinese astronomers was that of a supernova in 1054. The records, which note the event as the "guest star," mention that it remained bright for about a year before again becoming invisible. We know today that this was the supernova that created the Crab Nebula. Catalogued as SN 1054, the actual supernova occurred on July 4, 1054, with some records stating that the brightest of the outburst actually lasted closer to two years rather than one, and another record stating that the brightness lasted for 653 days, with 23 days when the supernova could be viewed during daylight hours.

The outburst occurred during the reign of Emperor Renzong of the Song dynasty, the year being recorded in Chinese documents as "the first year of the Zhihe era." The location of the guest star was cited "to the southeast of Tianguan (Zeta Tauri) perhaps several inches away." This has perplexed modern astronomers, because the Crab Nebula is not situated in the southeast, but to the northwest of the star Zeta Tauri, the constellation of Taurus. No fewer than three Chinese documents indicate that the guest star was located here, southeast of Tianguan. Much debate has followed in order to account for this discrepancy, allowing some to speculate that the supernova and Crab Nebula formation are not related.

There are references to the supernova of 1054 in a Japanese document discovered from the 13th century, and there is also a record of the supernova made by the Anasazi Indians of the American southwest. Ironically though, given the magnitude of such a sighting, there is no known record of the event in European or any other cultures. What few references there are from European sources appear doubtful in nature.

Situated in Taurus the Bull, the remnant of SN 1054, which consists of debris left by the explosion, makes for a worthy sight in any telescope. With the date of the supernova known, the Crab Nebula also makes for a rarity in the night sky. The nebula and associated pulsar contained within its cloud of dust and gas make for the two most luminous objects in their respective categories. At a distance of 6,500 light years (ly), the nebula spans 11 ly across and is thought to be growing at a rate of around one billion km per day, producing 75,000 times more energy than the Sun and emitting radiation from the entire range of the electromagnetic spectrum. The bright filaments or threads of the nebula have a temperature of around 27,000 °F.

The Crab Pulsar, discovered in 1968, is estimated to be 30 km in diameter, with at least 1.5 times the mass of the Sun. One of the few pulsars to be identified optically, this young neutron star rotates its "beams" every 33 milliseconds, or 30 times each second. Radiation is beamed out along the magnetic poles, and pulses of radiation are received as the beam crosses the Earth, much as a beam from a lighthouse causes a similar flashing mechanism.

Chapter 4

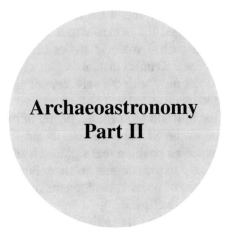

Archaeoastronomy
Part II

Neolithic Astronomy in Britain and Western Europe

The great stone circles across Europe define an era in astronomy. These huge monuments, many of which survive into the present, were built to help predict astronomical phenomena. Beginning around 3000 BCE, people across Britain and Western Europe began accumulating these giant stones known as megaliths, placing them in specific shapes and special orientations. The stones, along with others not just in Britain and Western Europe but also in North America and India, suggest that the marking of time was central to these early communities, who were developing increasingly sophisticated ways to measure the months and seasons. It can be assumed that these early constructions primarily served religious and ceremonial functions, with many attuned to the solstices for that reason.

The most famous example of all is Stonehenge in Wiltshire, England, which is aligned to the solstice, more than likely the winter solstice rather than the summer solstice. This is because those who worked the land would need to know when they should start preparing to plant the spring crop. The agricultural value of the stones to these local inhabitants is not doubted, but for us to believe that this was the stone arrangement's sole purpose seems somewhat naive.

© Springer Nature Switzerland AG 2018
J. Powell, *Rare Astronomical Sights and Sounds*, The Patrick Moore
Practical Astronomy Series, https://doi.org/10.1007/978-3-319-97701-0_4

Stonehenge is situated on Salisbury Plain two miles west of Amesbury and dates from between 3000–2000 BCE. It consists of standing stones, each around 13 feet high, seven feet wide, and weighing around 25 tons. It is thought that oxen dragged the stones across the land from a site up to 32 km away, while the central volcanic stones are thought to have been moved from Wales, some 210 miles distant.

The site is aligned not only with the solstice, but also with the Sun and the Moon. It is commonly accepted that Stonehenge recorded the rising and setting positions of the Sun and Moon at the height of each season. In addition, the oldest stone at the Stonehenge site, called the Heel Stone, was placed at the entry in such a position that sighting it from the center of the monument points directly to the summer solstice. Further, it has been suggested that an outer series of holes at the Stonehenge could have acted like a computer to predict lunar eclipses.

Over the years that followed its initial construction, Stonehenge was continually reconstructed, being used throughout this process of augmentation and recreation for astronomical purposes by the acting "custodian" of the time.

Studies conducted at sites throughout Britain and France indicate that many such structures have a mathematical significance as well.

Ancient Astronomy of the North American Indians

With such a wide and varied tribe culture across North America, identifying one particular tribe's engagement with astronomy against that of another tribe is difficult. So rich and diverse were the traditions and legends that it is hard to clearly separate astronomical involvement. It is easier to pick out the recurring elements among them. Early tribal society was one of more nomadic hunters and gatherers, contrasting sharply with other ancient cultures of the time, which developed the practice of astronomy after transitioning to an agriculturally based system.

The Anasazi

The Anasazi ("Ancient Ones") were thought to be the ancestors of the modern Pueblo Indians who inhabited the Four Corners region of southern Utah, southwestern Colorado, northwestern New Mexico, and northern Arizona, from around 200 to 1300 CE. Research traces the origins of the Anasazi back to people around 6000 BCE. From an early time, it appears that the Anasazi were typical of the aforementioned description—hunters, gatherers, and foragers who were somewhat nomadic in their wanderings. Not a great deal is known about the Anasazi, aside from these few general observations and their high cliff dwellings during a later, more settled phase, the remains of which can still be seen today.

However, the Anasazi did leave behind remarkable cave art at a site called Penasco Blanco ("White Bluff" in Spanish), which appears to show the night sky and what could be a supernova explosion. The artwork, which depicts a crescent moon alongside the representation of a bright star, could well point to the supernova of 1054. Made famous by the late astronomer and cosmologist Carl Sagan (1934–1996), this pictograph stands as a testament to how a culture lacking the accuracy of Chinese and another astronomers was still able to make an impressive, lasting mark on the overall foundations of archaeoastronomy.

Another such site is the Anasazi Sun Dagger, where a spiral design is traced into a cave wall. During midsummer, midwinter, and the equinoxes, it is accurately bisected or surrounded by daggers of sunlight that enter the specially placed windows. In another Anasazi ruin in Hovenweep National Monument near the borders of Utah and Colorado, these "light daggers" also illuminate spiral petroglyphs during the summer solstice.

Pueblo Indians

The Pueblo Indians, situated in the southwestern United States, are one of the oldest cultures in the nation. Their name is Spanish for "stone masonry village dweller." During their long history, it is thought that they evolved from nomadic hunter-gatherers into a sedentary culture, primarily making their homes in the Four Corners region. While hunting remained part of their culture, the Pueblo Indians expanded their knowledge of agriculture by growing maize, corn, squash and beans, all linked by a complex irrigation system. This settlement phase strengthened the society's cultural connections to the seasons and the annual rhythmic cycle of the planting and harvest times.

This agricultural lifestyle was interwoven with a strict religious community structure. The timing of ceremonies was critical. The Pueblo Indians devised a type of knotted cord that allowed them to keep track of solar cycles, with the summer solstice a particularly important marker in the year. A member of the Pueblo Indians known as a Sun priest was given the task of watching for the summer solstice through a notch in the wall of a specifically constructed "Sun tower." At the appropriate time, the priest would warn the people, speaking words to them that were thought to come directly from the Sun itself.

There was a strong belief that both sacred time (for religious purposes or ceremonies) and secular time (everyday, non-religious time) was regulated by the Sun, Moon, and stars. Thus, religious events would take place when the Sun, Moon, and stars were all in the correct position. The Sun priest was also responsible for determining the exact time, accomplished by looking at the Sun over the horizon. Accuracy is debatable but can be considered reasonably good, depending on what the tribe required the time for. There was of course some margin of error with regard to giving specific times of the day, but this would not have been as important to the Pueblo Indians, as their most important cycles weren't governed by such precise accuracy.

Pawnee Indians

The Pawnee Indians were known to have a complex religion, a big part of which was centered on astronomy. Considered to be among the best sky watchers of ancient times, the Pawnee were especially interested in the gods of the sky, with the North Star venerated as the beneficent creator god and the South Star seen as a magical force of opposition belonging to the underworld. The Pawnee Indians entrusted much in the stars, with their most sacred ceremonies and measurements of time tied to the movement in the skies. They developed a series of complex star rituals and ceremonies, with their celestial religion considered to be the most sophisticated one recorded for a North American culture at this time.

Not only were the Pawnee au fait with the stars, but records also show their familiarity with meteorites, with the development of their own observing methods that followed the track of the Sun, Moon, and five planets. The Pawnee believed that they were once born of the stars and were able to map their favored stars on buckskin.

The Skidi band of Pawnee Indians were not closely bound to other Pawnee bands of Indians, and there is a possibility that their ancestral lines were different. Their sky map gives no indication of the importance of the path of the Sun, the ecliptic, or the summer solstice. They also seem to have lacked any precise lunar calendars. Collectively, it may well be the case that the Pawnee Indians did not consider the solstices to be of any importance.

However, in other aspects, the Skidi Pawnee, or the Wolf Pawnee, did go to great lengths to feel connected with the sky. As a running theme throughout the Pawnee as a whole, one particular Skidi buckskin chart depicts some particularly important stars that guided the Skidi in their lives. While the Pawnee tribes erected their small houses and villages following constellation patterns observed with the naked eye, the dominance of such connections in the construction of the Skidi villages is even more significant. The Skidi village designs were laid out in the position of the most important stars in the sky, with the last corner of the village dedicated as a shrine to the morning star (Venus). Similarly, to the west, there was another shrine dedicated to the evening star. The doors to the houses (lodges) always faced east to greet the rising Sun, with four painted poles representing the four important semi-cardinal directions: northwest, northeast, southwest, and southeast. The Pawnee collectively believed that the positions of the poles, which were used to hold up the structure, represented the support given to the sky by the star gods who dwelled in each of the four semi-cardinal directions. Gods of the four cardinal directions, north, south, east, and west, were also of great importance. The domed roof on top of the construction represented the sky.

One particularly wonderful piece of Pawnee Indian cosmological myth describes the creation of the universe, referring to Mars as the "Red Morning Star Warrior" and Venus as the female "Evening Star." In Pawnee writings, Mars mated with Venus to produce a daughter, who subsequently gave birth to the first humans that would populate Earth. Man was created by the overlapping of the Sun and the Moon. The Pawnee believed that when the end of civilization was coming, they would be warned by the Moon changing to a darker color and then to black. When this cataclysm took place, the Sun would dim quite quickly and then suddenly all would be dark, much darker than an eclipse.

Chumash Indians

The Chumash Indians' homelands lie along the coast of California between Malibu and Paso Robles, as well as on the Northern Channel Islands. The peoples there are known to have resided in the area for thousands of years, with numerous archaeological sites uncovering a history that dates back 15,000 years. Like the Pawnee Indians, the sky for the Chumash significantly impacted their lives, so much so that each object in the sky held a specific rank of importance.

The Chumash also held the objects in the sky as supernatural gods. They placed the Sun, an aged widower who carried a torch through the sky, at the top of the rankings. Next was the Moon, a female god who controlled human health. Venus was viewed differently, depending on where the planet was sighted: as a morning star, a kind god; as an evening star, an evil god representing the underworld. The Chumash held the time just before the winter solstice as their most important time of the year, for they believed that this was when the Sun might decide not to come back to the Earth.

In order to ensure the Sun's return, prayers were held along with dancing in ceremonies that would last for several days. Without the Sun, the Chumash knew there would be no life for them. The praying and chanting would continue as the Chumash tried to pull the Sun back to Earth.

The discovery that the Chumash used tally cords and notched sticks to record astronomical records is a clear indication of their close links with the workings of the night sky, with a great many stars having been recorded right down to sixth magnitude in brightness. Records show that beyond naming many of the stars, they also identified Mars, Jupiter, and Saturn.

The Big Horn Medicine Wheel, Wyoming

One of the great mysteries that still evokes debate is that of the medicine wheel, one of which is located in Wyoming. Formally known as the Big Horn Medicine Wheel, the renamed Medicine Wheel/Medicine Mountain National Historic Landmark is situated in the Bighorn National Forest and is comprised of a network of large stones sourced from local white limestone laid upon limestone bedrock.

The stones are positioned at the summit of Medicine Mountain, 2,939 meters high, with its construction by Cheyenne Indians dating back between 300–800 years. Some question this analysis, as there were many ancient Plains Indian tribes who led nomadic lives across central North America. Local tribes interviewed at the time of its discovery in the 19th century reported it was already there when they arrived, describing the Big Horn Medicine Wheel as coming from the "people who had no iron."

Archaeologist John H. Brumley developed an official definition for the medicine wheel. According to Brumley, a medicine wheel contains at least two of the following three criteria:

1. A central stone cairn;
2. One or more concentric stone circles; and/or
3. Two or more stone lines radiating outward from a central point

By Brumley's definition, the number of remaining medicine wheels is estimated at between 100 and 200, each of which, while fitting the criteria, has a unique form. The vast majority of the medicine wheels are to be found in Alberta and Saskatchewan, though some are located in North Dakota, Wyoming, Montana, and Colorado. These medicine wheels, or "sacred hoops" as referred to by some, pose many unanswered questions about their origin and purpose. In general, the wheels are thought to provide a multitude of uses to various tribes.

The Big Horn Medicine Wheel in Wyoming consists of a circular rim measuring 24.3 meters in diameter, with 28 spokes extending from the rim to the center, plus a series of seven stone circles, known as cairns. At the center of the structure is Cairn O, measuring three meters in diameter. The other cairns, lettered A to F, are positioned at or near to the rim and are noticeably smaller in size. Cairn O is thought to represent the oldest part of the Big Horn Medicine Wheel, with excavations conducted in the area showing that Cairn O actually extends below the wheel, being covered over the years by dust blown across the wheel by the wind. It is quite possible that it originally supported a central pole.

Most of the A to F cairns are large enough to hold a sitting human, and if you stand or sit at one cairn facing another, you will be pointed to certain places on the distant horizon. These points indicate where the Sun rises or sets on summer solstice and where certain important stars rise heliacally (in sequence), the first to rise at dawn after being behind the Sun. It is believed that the dawn stars were crucial in helping foretell when the Sun ceremonial days would be coming. The star alignments are most accurate for the period around 1200 CE, since slight changes in the Earth's orbit have caused perturbations since then. However, the solstice alignments remain accurate in the present day. As the structure is generally covered in snow for much of the year, it is only around the time of the summer solstice, a window of two months, that the Big Horn Medicine Wheel can be properly reached and viewed.

The 28 spokes of the Big Horn Medicine Wheel correspond with the same number used in the roofs of ceremonial buildings, such as the Lakota Sundance Lodge. Lodges like this one always include an entrance to the east, facing the rising Sun, and 28 rafters for the 28 days in the lunar cycle. The number 28 itself is significant to a number of Indian tribes because of its relation to the period of one lunar month. It is thought that, in the case of the Big Horn Medicine Wheel, the number 28 could refer to the dawn rising of Rigel in Orion the Hunter, 28 days past the summer solstice, and also Sirius in Canis Major, rising another 28 days past the rise of Rigel. The stars in the constellation of Orion and Canis Major, most notably Rigel and Sirius, have often been found aligned with other sacred sites around the world, all connected to the religion of the Sun. The Moai statues on Easter Island face the heliacal setting of Orion; similarly, the ancient stone circle at Nabta Playa in Egypt demonstrates heliacal alignments with Orion.

A fair proportion of the work conducted at the Big Horn Medicine Wheel was carried out by archaeoastronomer John Allen (Jack) Eddy (1931–2009). Eddy, who published professionally under the name John A. Eddy, visited the site in the early 1970s, making an in-depth study of the arrangements of the rocks, cairns, and spokes. Eddy discovered the alignments to the rising and setting of the Sun at summer solstice and to the rising places of not just Rigel and Sirius, but also Aldebaran in Taurus—collectively, all bright and significant stars associated with the solstice. Eddy noted that Cairn E and the middle Cairn O perfectly aligned with the direction of the summer solstice sunrise. Standing at Cairn E and facing Cairn O, you would see the summer solstice sun rise up on the horizon, aligning the central cairn with Cairn E. This was also the case for the setting of the summer solstice sun, again perfectly aligned with Cairn A and Cairn O.

Eddy also documented that from the vantage point of Cairn C, the sun would set in alignment with C and the central mound. He also noted that, whilst standing at Cairn F two days before the summer solstice, Aldebaran would rise in alignment with Cairn A, visible for a short time as it emerged in the sky just before dawn. Rigel's appearance 28 days after the summer solstice saw this bright star rise just above the Sun, in alignment with Cairn F and Cairn B. 28 days after, Sirius rose in alignment with Cairns F and C.

Eddy's work was added to in the late 1970s by astronomer Jack H. Robinson (1925–2016), who found a cairn pair that marked the bright star Fomalhaut's rising point with the Sun, 28 days before the solstice.

The Majorville Medicine Wheel

Beyond the Big Horn Medicine Wheel, the second largest extant medicine wheel is located near Majorville in southern Alberta. The Majorville Medicine Wheel consists of a central cairn measuring nine meters across, with 28 spokes radiating outward to form an outer ring that is 27 meters in diameter. Thought to date back to around 5,000 years ago, the Majorville Medicine Wheel is not only the second largest but alsothe oldest known wheel. Archaeologists arrived at the 5,000-year dating by examining the styles of spear points discovered at the site, with the changes in point styles reflecting different time periods. This method, while widely accepted, was still subject to challenge. The time period was later corroborated by precise radiocarbon dating conducted on a bone fragment found at the bottom of the cairn. Scientist Gordon Freeman asserts that the Majorville Medicine Wheel stones are in fact the remains of an open-air Sun temple that predates both Stonehenge in England and the pyramids in Egypt. Freeman also asserts that the Plains Indians of the time used the temple to observe sunrise on the winter and summer solstices. Remarkably, Freeman notes that that Britons and Plains Indians made precise astronomical observations at these two sites halfway around the world from each other at nearly the same latitude.

Moose Mountain Medicine Wheel

In Saskatchewan, the Moose Mountain Medicine Wheel commands a different geometry to that of the Big Horn Medicine Wheel and the Majorville Medicine Wheel. Situated at Moose Mountain, the central cairn is surrounded by just five spokes, not 28, with an outer ring nine meters in diameter. The spokes extend far beyond the outer ring, with the longest spoke measuring 37 meters. This spoke marks the summer solstice sunrise. Still, Moose Mountain Medicine Wheel does have cairn alignments that point to the heliacal risings of Aldebaran, Fomalhaut, Rigel, and Sirius.

The Moose Mountain Medicine Wheel has been carbon dated to around 2,600 years old, first appearing in the records of surveyors who documented the area in 1895. The surveyor records refer to a central cairn measuring about four to five meters high, although in the modern day, the cairn measures just over half a meter high. The explanation for this shrinkage has been put down to vandalism and theft, something all medicine wheel sites have sadly been vulnerable to over the years.

At the Moose Mountain Medicine Wheel site, archaeologist Geoffrey Ian Brace challenged the accepted idea that the wheel was used for astronomical usage. Brace undermined the astronomical importance by claiming that carrying out alignments over the six-meter crest that separates one end of the Moose Mountain Medicine Wheel from the other would be far too difficult. However, John Eddy calculated the probability of a chance of alignment to be less than one in 4,000.

The Incas

The Inca civilization in Peru was a powerful empire that dominated for a century, being halted at last by the Spanish conquest of the New World. This well-organized, well-disciplined society was brought together by Pachacuti Inca Yupanqui, a military leader who managed to bring South America under one rule following the demise of the earlier Huari and Tiwanaku cultures. The center of the empire operated from the city of Cuzco in the Andes Mountains, commanding a vast swathe of land that encompassed 605,000 km^2. The Inca domain stretched from present-day Columbia to Chile.

Astronomy played an important role in the society of the Incas, primarily due to the importance of agriculture. Cuzco, right at the heart of the empire, was seen as the very center of affairs, the political and spiritual focal point of the Incas. The city itself was laid out in a radial plan that mimicked the sky and pointed to specific astronomical events on the horizon. As with the Ancient Indian and Egyptian cultures, the most important events involved the risings and settings of the Sun, Moon, and stars. This horizon-based culture had a great effect on the rulers and people of the time.

This strong link is exemplified by the emergence of the Pleiades star cluster above the horizon. As the cluster rose, it marked the beginning of the new Incan year. While we also know the Pleiades star cluster as the Seven Sisters or Seven Kids after the seven brightest stars in the cluster, it has been recorded that the Incas, because of their clear atmosphere at the high-altitude city of Cuzco, actually made out 13 stars!

The use of astronomy was chiefly dedicated to agriculture. The Incas carefully erected stone pillars on the foothills overlooking Cuzco to match specific orientations. When the Sun rose or set between the pillars, it was time to plant at specific altitudes. These pillars formed a massive timepiece, marking time as accurately as possible for the high altitudes right down to the floor of the valley, and everywhere in between. The Incas made sacrifices to the Sun in the hope that the Sun would rise in the proper place for planting.

Astronomers recognized that the planet Venus was in fact the same planet, whether it rose in the morning or was visible in the evening. They believed that Venus was a servant of the Sun and was ordered to either go ahead or stay behind the Sun, but to always remain close. The Incas took great interest in other stars and the patterns they formed. They spotted many animals in star formations, along with other symbols from their everyday lives. They believed that the Viracocha, the supreme god of the Incas, had ensured that every animal or bird had a corresponding star and that all living things would be protected.

The Incas sorted the constellations into two groups. The first and most common group contained the stars that were joined in a dot-to-dot fashion, linking together to form pictures of animals and gods, among others. The Incas considered this grouping to be inanimate.

The second group of constellations could only be observed when there were no stars visible, singling out dark spots and "blotches" on the Milky Way. The Incas believed these blotches to be animate representations of animals living in certain pockets of the Milky Way, which itself was thought to be one great river. This rather creative innovation made the Incas one of only a very few civilizations to do so, stemming from their deeply held belief that everything around them was in some way connected.

The chief observatory of the Incas was called the Coricancha, or "golden enclosure," and was covered completely in gold. A gold disk faced the rising Sun. Coricancha, or "temple of the Sun," was situated in the heartland of the

Incan Empire in Cuzco, the city itself said to be designed in the shape of a puma, with Coricancha located in the animal's tail. Coricancha was placed at the convergence of four main highways and was directly connected to the four districts of the empire. Housing in excess of 4,000 priests, the temple cemented the symbolic importance of the Incas' religion, uniting the divergent cultural practices that were observed throughout the vast territory.

The positioning of the temple in relation to the nearby Andes meant that Coricancha functioned as an enormous calendar: the shadows cast by strategically placed stone pillars in the foothills could be viewed from the temple, marking out the solstice and equinoxes. As previously discussed, these stones were also part of the great timepiece that predominantly served agricultural purposes.

The temple consisted of four main chambers, each dedicated to a different deity—that of the Moon, stars, thunder, and rainbows. Much of the Coricancha was filled with gold, with the giant disk that faced the Sun reflecting sunlight to illuminate the rest of the temple. The disk also served a more intricate and intriguing purpose. During the summer solstice, it illuminated a sacred space where only the emperor, Pachacuti, was allowed to sit.

Sadly, the glory of the empire was shortened following disputes over successions, the invasion of the Spanish in the 1530s, and a deadly smallpox epidemic. Coricancha was all but destroyed, its gold melted down and sent back to Spain. Many centuries later, an earthquake ruined the Spanish-made cathedral that stood where Coricancha once stood.

Maya Astronomy

The Maya civilization began at about 500 BCE, succeeding that of the Olmec Empire. Over time, the influence of the Maya civilization spread to cover much of Central America, including what we now know as Mexico, Guatemala, Belize, and Honduras.

Maya society considered astronomy to be extremely important, as the Mesoamerican culture believed that the science and its study reflected the great order of the universe in which they lived. Astronomy gave a corresponding order to them and the gods that they worshipped. The order inherently fused their theological views with the physical world around them.

The capturing, interpretation, and overall accurate measurement of time was an important factor for the Maya, for in their cosmology and overall understanding of the universal model, space and time were inseparable. This binding is evident in the complex calendar system perfected by the Maya, a system so collective in its construction that spatial attributes of the universe, such as plants and animals, were interconnected with the temporal movements of astronomical objects. For the Maya, the sky was their method of accurately marking the passage of time.

The Maya civilization developed the hieroglyphic script, the only known fully developed writing system of the pre-Columbian Americas. The script is thought to have been inherited from the Olmec, a culture that is generally regarded as the originator of all of the great Mesoamerican cultures. It has also been documented as showing great mathematical skills for the time period, introducing the concept of zero, which was very rare in Eurasian cultures at that time. Coupled with the Maya's insights into astronomy, art, and architecture, it is safe to say that this was a well-balanced, impressive society.

The Maya believed that celestial events indicated communication with the gods, with certain astronomical objects representing certain deities, whose lives were portrayed in the daily, monthly, and annual changes in their appearance. Religion moved astronomy into astrological territory. The seasonal movement of constellations across the night sky, along with celestial events, was thought to have a direct effect upon the lives of the Maya people.

As with other cultures, astronomy's most practical benefit was in the realm of agriculture, with the appearance or disappearance of certain constellations or planets heralding the time for planting and harvesting of the crops.

It is sometimes argued that the Maya's use of astronomy wasn't for scientific purposes at all, and that astronomy was not in fact used to mark the passing of seasons for agricultural production. One school of thought places astronomy at the feet of the Maya priesthood alone, being solely used as a tool to understand the past cycles of time and to prophesize about the future.

The Maya priest-astronomers adopted the role of *ilhuica tlamatilizmatini*, or "wise man who studies the heavens." These men commanded great power, given the belief that they could actually predict the future. The priesthood refined their observations, making careful note of eclipses of both the Sun and the Moon, along with the movements of stars and that of Venus. These observations were then measured against similar events recorded in the past, on the assumption that similar events would occur in the future when the same astronomical conditions prevailed. Their knowledge of the patterns of the sky pushed the priests to an exalted position in Maya society, with priest-astronomers spending many dark hours studying the nighttime sky and the exact time. As the length of day varied with the passage of seasons, they had to be very knowledgeable about the sky in order to measure down to the precise hour, and then to subsequently predict when the Sun would rise again.

The priest-astronomers declared the commencement of a new day to be at sunrise, which was at odds with some Maya. The alternate belief was that the day officially started at noon, when the sun was at its highest point in the sky, or even at sunset! This alternate belief may seem quite bizarre to the conventions of the Western world, but it does have credence in its premise. In Western society, we start the day when the Sun is almost at its anti-zenith, when it is at its highest point on exactly the other side of the Earth.

The priest-astronomers recorded Maya cosmology in codices, some of which remain despite the Spanish burning a substantial amount of them. Illustrations in the codices show that priest-astronomers made astronomical observations with the naked eye, assisted by crossed sticks as a sighting device, a sort of view-finder. In what remains of the Post-classic codices, the Maya society recorded eclipse tables and made calendars with astronomical data and a level of accuracy superior to that of the accumulated knowledge in Europe at this time.

The Maya had two main calendars. The first was the ceremonial *Tzolk'in*, a 260-day calendar of 13 numbers and 20-day names. The second was a vaguer calendar, *The Haab*, made up of 365 days. *The Haab* calendar had 18 months consisting of 20 days, with a five-day month added at the end of the year. The 20-day month is largely based upon the Maya's vigesimal number system, which is a base 20 system, as opposed to our base ten decimal system. Evidence suggests that the Maya were well aware that there were 365 days in a year, but beyond the fact that it did not fit in with the vigesimal system, it was probably far easier to keep the calendar the way it was rather than cause massive disruption by instigating such a society-wide change.

One of the codices to survive the Spanish invasion, named Codex Vaticanus A, imparts a wealth of knowledge about the Maya concept of the universe. This document alone portrays the Maya belief in a multi-layered Universe consisting of 13 levels of the heavens and nine layers of the underworld, with the Earth sandwiched in between and considered to be a part of both. The Milky Way was seen as a road of souls traveling to the underworld, or as the umbilical cord connecting heaven and the underworld to Earth. Each level had four cardinal directions associated with a different color. Major deities had aspects associated with these directions and colors: north was white, east was red, south was yellow, and west was black.

While appreciating the vast expanse of the night sky, the Maya seemed to focus their attention on several objects in particular: the Sun, the Moon, and Venus, with a handful of constellations and stars and the Pleiades star cluster making up their collective studies. Priest-astronomers spent generations determining and recording precise paths of these celestial objects as they moved through the night sky and through the seasons.

The most important object for the Maya was the Sun, which was universally recognized as the life-giver on Earth. The Maya associated the god *Ilhuicati Tonatiuh*, a red eagle with a large and all-seeing eye, with the Sun. Their observations made of the Sun are quite astounding, given that they had to account for the tilt of the Earth's axis and the corresponding positions that the Sun would therefore be seen season to season. Study of their observations reveals a great accuracy in the calculated times for the Sun to rise and set, with records showing further calculations of 365 days to represent a solar year. A tropical year is actually 365.2422 days long, an inaccuracy that would have grown more noticeable over time. The priest-astronomers seemed aware of this, with evidence suggesting that they continually updated and amended subsequent records.

Solar and lunar eclipses were considered by the Maya to be dangerous events that would bring catastrophe upon the world. Another of the codices, the *Dresden Codex,* the oldest known book written in the Americas, depicts a solar eclipse through a serpent devouring the *k'in* (day) hieroglyph. Eclipses in Maya culture were generally thought to represent the Sun or the Moon being bitten. Lunar tables recorded eclipses so that the Mayas could predict future events and therefore have time to ward off the impending disaster by performing certain ceremonies. The *Dresden Codex* was successfully smuggled out of Central America at a time when the Spanish destroyed many Maya documents, considering them pagan. Of the many codices that are known to have existed, only four remain, all named after the European cities where they are kept. The three European codices are named after Dresden, Paris, and Madrid, with the fourth disputed book, the Grolier Codex or Grolier Fragment, residing in Mexico City. Despite the efforts in the mid-16th century by the Franciscan missionaries to burn to eradicate the religion, these three, possibly four religious artifacts have been saved.

The *Dresden Codex*, made of ficus tree bark and covered with lime for a glossy finish, arrived at its home in the Royal Library in Dresden in 1739 after being purchased from a private collector in Vienna. It is believed that the codex was the center of a daring rescue from the Yucatan Peninsula as other codices met their fate at the hands of the missionaries. The codex was drawn by no fewer than eight different scribes, and it is believed that it was created sometime between 1000 and 1200 CE during the Post-classic Maya period. This particular codex deals primarily with astronomy, calendars, and good days for rituals, planting, and prophecies. There are also references to sickness and medicine.

The *Paris Codex*, which was rediscovered by chance in 1859 in a dusty and neglected corner of the Paris library, is not as complete as the *Dresden Codex*, but nevertheless does contain fragments of 11 double-sided pages. This codex is believed to date from the late Classic or possibly Post-classic era of Maya history. It covered astronomy, along with Maya ceremonies, dates, historical information, and descriptions of the Maya gods and spirits.

The third European-based codex, the *Madrid Codex*, was separated into two parts after it reached Europe and was initially accepted as being two completely different codices. In 1888, the document was put back together to form the codex as it is known today. As many as nine different scribes are thought to have contributed to the text, with astronomy, astrology, and divination the main topics covered. Despite its poorly drawn appearance, the *Madrid Codex* is of great importance for its content on Maya gods and the rituals associated with the Maya New Year. Information on different days of the year is documented, detailing the particular god associated with each. The codex also includes an interesting insight into Maya activities, including hunting and pottery making.

The *Grolier Codex* is incomplete, thought to be a fragment of a much larger text than its present 11 rather battered pages. Astrology and the movements of Venus form part of the text, and despite lingering doubt over its authenticity, many historians now accept the *Grolier Codex* as genuine. This *Grolier Codex* (sometimes known as the *Saenz Codex*) gained international attention in 1971, when it was displayed at the Grolier Club in New York (America's oldest society for book and graphic art) in an exhibition entitled "Ancient Maya Calligraphy." The codex, which consists of fig bark sheets stuccoed and painted on one side, is said to have been originally recovered from a cave in the Mexican state of Chiapas in the 1960s, making its appearance at the Grolier Club from April 5 to June 5, 1971.

Beyond the crossed sticks, the Maya had little in the way of observational equipment for astronomy. Most all of their work was completed with the naked eye. Unlike other civilizations, there is no record of any possible armillary spheres or sextants. Despite this apparent disadvantage, the Maya excelled in the area of design and construction, with many of the temples erected and aligned to help observers monitor celestial positions. Many of the temples pointed towards the equinoxes or summer solstice, while other buildings had doorways and windows aligned with the most northerly or southerly risings of Venus.

Venus held a particular attraction to the Maya civilization, considered to be connected with the major deity Quetzalcoatl. Venus was called Xux Ek, meaning the "Great Star," with the Maya able to identify that this was the same object seen both in the morning and evening sky. A fair proportion of knowledge about the Maya and their attraction to Venus lies within the 30 double-sided pages of the aforementioned *Dresden Codex*. The text of the codex reveals laborious measurements of the rising and setting of the planet Venus. The rising of Venus in the mornings was considered bad luck, and everyone would stay inside their homes and block their chimneys so that the evil light from Venus could not enter.

The priest-astronomers determined the period Venus takes to orbit the Sun to be 584 days. They counted five sets of 584 days, (that is 2,920 days is approximately eight years or five repetitions of the Venus cycle). Their calculation of Venus' orbit is incredibly accurate, given the actual synodic period of 583.92 days. Some findings suggest that the Maya considered Venus more important than the Sun, with records showing that they made detailed daytime observations of the planet and hinting at the great psychological effect it had not only on the Maya but on other Mesoamerican cultures as well. The waging of wars by the Maya was timed based on the stationary points of both Venus and Jupiter. Human sacrifices occurred on the first appearance after Superior Conjunction, when Venus was at its dimmest magnitude. The Maya most feared the Heliacal Rising after Inferior Conjunction.

Beyond the calculation of the synodic period of Venus, the Maya calculated the period for Mercury at 117 days (actually, 116 days) and also for Mars, at 780 days (actually, 779.936 days). Records confirm that the Maya recognized the existence of both Jupiter and Saturn, but no documentation exists of any similar orbital computations for those planets. Strangely, none of the planets catalogued were acknowledged as more than stars themselves, despite their unique movement against a backdrop of fixed star positions.

The Moon too held great significance to the Maya, represented by a female deity who had a powerful influence on terrestrial events. Maya dynasties often claimed to be descended from the Moon. A waxing Moon had the attributes of a beautiful woman, whilst a waning Moon was considered to be an old female who ruled over childbirth. Documentation relating to calculations of a lunar cycle has been discovered, with one Maya astronomer said to have calculated that there were exactly 149 moons over a period of 4,400 days, which works out to be an average lunation of 29.53 days. In the city of Palenque, also known as Lakamha, a Maya city-state that stood in southern Mexico, astronomers found that there were 405 moons in 11,960 days, providing an average lunation of 29.53086 days. The degree of accuracy is striking—the actual average lunation is 29.53059 days.

Certain star clusters were known to have a special place in Maya culture, probably none more so than the Pleiades. Its morning appearance during April signaled that planting should commence. The Maya made note of the constellations around the Pleiades star cluster, so that the seasonal change aside from the Pleiades rising could be captured and understood over the years. So high was the regard for the Pleiades cluster that the culture named it Tianquiztli, meaning "marketplace," believing that the cluster was at the center of the layer of fixed stars (rather than Polaris the Pole Star), around which the rest of the sky seemed to revolve. The Maya believed the Solar

System rotated around Alcyone, the central star in the cluster, in a 26,000-year period. The impressive city of Teotihuacan was erected by its ancient builders so that the main street was aligned with the Pleiades.

Naturally, a sky that is closer to the equator allows the observer to see constellations that could not be seen if positioned further to the north or south—indeed, exactly twice the number of constellations that can be viewed from the poles. This vantage point allowed the Maya to observe such constellations as the Big Dipper, Orion and in particular Orion's Belt, and Cassiopeia. They believed these prominent star groupings had a vital significance in the overall makeup of the cosmos. Festivals were held when the Pleiades or Orion's Belt rose at sundown, then vanished with the dawn of a new day.

Aside from the Sun, Moon, stars, and constellations, it has been documented that the appearance of a comet, a "star that smokes," foretold the death of a noble in Maya society. Such links between celestial and terrestrial events became deeply embedded into the Maya belief system over time.

We know that the highly regarded priest-astronomers made predictions of when certain astronomical objects would be seen throughout the year. This was done through continued study of the night sky as it changed over the seasons. However, for certain astronomical events, mathematics was required. The Maya developed the most sophisticated mathematical system in the Americas. The system required only three symbols, a dot representing a value of one, a bar representing five, and a shell representing zero. These symbols were used in various combinations to keep track of calendar events both past and future, a system so precise and yet so seemingly simple that even uneducated citizens could engage with the simple arithmetic required for trade and commerce.

Several numbers were considered to be of particular importance to the Maya. The first was 20, which represented the number of fingers and toes a human being could count on. The number five was also important, corresponding to the number of digits on a hand or foot. 13 was sacred, as it represented the number of Maya gods. 52 represented the number of years that were referred to by the Maya as a "bundle," a unit concept similar to that of a century. 400 had another sacred meaning, representing the number of Maya gods of the night.

The vigesimal system (base 20) is actually far simpler than that of the Roman numerals system. Despite its lack of fractions, some amazingly accurate results were achieved through its use. The Maya took on far more challenging tasks than predicting when the Sun would rise and set, such as predicting solar and lunar eclipses. As previously discussed, solar eclipses, known as *chi' ibal kin*, or "to eat the Sun," were viewed as truly terrifying events that caused widespread panic.

The predicting of eclipses required close study of the Earth's movements in relation to those of the Sun and the Moon, and the calculations would need to consider orbital plane differences. The Maya did not know that the Sun is both 400 times larger than the Moon and 400 times farther away. Despite that, pre-Columbian expert anthropologist duo Harvey Miller Bricker (1940–2017) and Victoria Bricker (1940–) made a startling discovery while decoding early Mayan hieroglyphics, that an astronomical calendar dating to the 11th or 12th century accurately predicted a solar eclipse to within a day in 1991, centuries after the Maya civilization had ended. The eclipse occurred on July 11, 1991, just as the Maya recorded it would.

Priest-astronomers determined the nodes when the paths of the Moon and Sun cross, every 173.31 days. It was observed that during this time, eclipses may occur within 18 days of the node. In the eclipse section of the *Dresden Codex*, two numbers appear frequently: 177, approximately the length of six lunations, and 148, the length of five lunations. The numbers 177 and 148 are representative of the times when eclipses were predicted. Occasionally, the number 178 would appear, as astronomers made periodic corrections and adjustments to their eclipse tables to account for minor errors in calculations. The tables, which consisted of columns and rows of numbers, utilized an eclipse glyph, symbolizing days when calculations predicted an eclipse. Should an eclipse occur, the number was replaced by the eclipse glyph.

Aztec Astronomy

For the Aztecs, astronomy was a study that was closely associated with religious significance and a strong moral code of behavior. Aztec astronomy played a significant role in society, forming a strong bond between the culture and the heavens.

Unlike other civilizations such as the Maya, the Aztecs seriously considered the possibility that the world could be destroyed and recreated at the end of a 52-year cycle, a complex calendar cycle used by the Aztecs and characteristic of Mesoamerican societies. Combining a count of 365 days based on a solar year, called xiuhpohualli (year count), with a separate calendar of 260 days based on various rituals called tonalpohualli (day count), every 52 years (called a calendar round) would see the calendars overlap and a new cycle commence. For many cultures, observing the night sky and recording the movement of various objects instilled a sense of uniformity, regularity, and harmony with the overall balanced cosmos. Not for the Aztecs, though, as their culture seemed to focus on the lack of stability and the potential destruction of all around them, with the threat of obliteration coming from the sky. This concept of the cosmos had far-reaching social consequences on the Aztec society.

The Aztec calendar had 20 days in a month, counting 18 total months. Since the calendar was tied to the solar cycle, it added up to a total of 365 days. Each 20-day month began with celebrations at a festival. Moreover, symbols were ascribed to each of the 20-day periods in the 365-day cycle. The 365-day cycle was marked by 360 named days and five nameless days. The Aztecs considered these last five days to be unlucky.

The 260-day "Ritual Cycle" was divided into 20 periods of 13 days each. Each cycle was associated with a specific Aztec deity; the second week, for example, was associated with Quetzalcoatl, one of the chief Aztec deities. This unusual approach meant that for the Aztecs, the movements of heavenly bodies and life on Earth exhibited a powerful bond. Not only were the Aztec cities designed, planned, and built in accordance with astronomical alignments, but the original location of a city itself was also determined by its potential significance for rituals guided by observations of the night sky.

The Aztec calendar to say the least was sophisticated, measuring the passage of time with an interconnected triple system of calendars that adhered to movements of celestial bodies, providing a comprehensive list of important religious festivals and sacred dates with a unique combination of names and numbers. In addition to the name and number, each day was given its own deity.

Other research contradicts this clear demarcation of days, arguing that each day contained different, potentially conflicting meanings, drawn from the complexity of the Aztec calendar. The calendar was an expression of affairs in which different deities could act in contradicting ways on separate days. The conflicting interpretations of such actions could severely hamper the Aztec society and lead to widespread confusion.

At the heart of Aztec cosmology stood the Sun, famously represented on the Aztec Calendar Stone. The stone provides a detailed depiction of the Aztec notions of time and space. It is divided into five portions, or five eras, which are associated with five different Suns. This follows Aztec mythology, which says that mankind perished at the end of the four previous destructive Sun cycles, and that we are currently living in the fifth cycle of the Sun. This fifth Sun had come into existence as a result of a god sacrificing himself for humanity. This earlier sacrifice mandated continued human sacrifice to the Sun and encouraged Aztec citizens to live lives that would prevent the gods from destroying the whole world again.

Over the years, many researchers have attempted to map the Aztec dates and cycles in terms of the Georgian calendar. It has been argued that February 23 marked the first day of the Aztec calendar. Correspondingly, the date of the Fire Ceremony was February 22, a day before the beginning of the New Year. However, there remains much debate over the so-called Legend of the Fifth Sun. Several different versions of this myth exist, as the stories were originally passed down by oral tradition, and also because the

Aztecs adopted and modified gods and myths from other tribes that they met and conquered. According to the general Aztec creation myth, the world at the time of the Spanish colonization was during the fifth era of a cycle of creation and destruction. During the 16th century and extending in a time frame that encompassed modern times, the Aztecs believed that humans were living under the fifth Sun, which would end in violence at the conclusion of the calendrical cycle—violence taking the form of earthquakes, and all people being eaten by sky monsters!

The Five Sun Cycles

1. Nahui-Ocelotl (Jaguar Sun)—Inhabitants were giants who were devoured by jaguars. The world was destroyed. The world lasted 676 years, or 13 52-year cycles.
2. Nahui-Ehecatl (Wind Sun)—Surviving inhabitants fled to the tops of trees and were transformed into monkeys. This world was destroyed by hurricanes. It lasted 676 years, or 13 52-year cycles.
3. Nahui-Quiahuitl (Rain Sun)—Inhabitants were destroyed by a rain of fire. In a majority of cases, only birds survived (or inhabitants survived by becoming birds). Further research reveals that the Aztecs though the survivors also became turkeys, butterflies, or dogs! This world last 364 years, or seven 52-year cycles.
4. Nahui-Atl (Water Sun)—This world was flooded, turning the inhabitants into fish. It lasted 676 years, or 13 52-year cycles.
5. Nahu-Ollin (Earthquake Sun)—We are the inhabitants of this world. This world will be destroyed by earthquakes (or one large earthquake).

Looking Forward

Understanding how our ancestors viewed the universe remains of great significance in the modern-day world. Firstly, it serves as a great reminder that what we presently perceive as being the true understanding of the universe may well be challenged and eventually tossed aside in the coming centuries.

We work with the evidence that we are given, and our ancestors did so to the best of their abilities, making observations, calculations, and judgements that were to ultimately shape the modern world of astronomy. Archaeoastronomy offers us a necessary insight into our past, so that the astronomers of the present can continue to shape the fundamental basics that define our existence in the universe.

Chapter 5

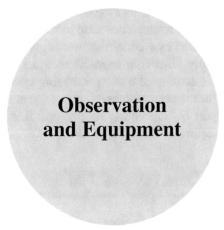

Observation and Equipment

The natural companion of the astronomer would most obviously be the telescope. Since its invention, the telescope and its association with this field strike a chord across the globe. However, as seen in the previous chapter, we must not forget other tools at our disposal: binoculars, and perhaps the most priceless instrument of all, our own eyes.

The ownership and use of a telescope or a pair of binoculars in the world of astronomy should not hinder the desire to enjoy astronomy for what it really is—a subject that everyone can enjoy and feel a part of. Nobody should be excluded from following their passion in this community because they lack such optical aids. Many early observations were made with the naked eye alone, and if it were not for these documented sightings, our understanding of the universe would be far less advanced than it currently is, which is why referencing our ancestors and their observations is so important. Yes, a telescope is of benefit and yes, more can be achieved through it, but at no point should anyone be discouraged otherwise if they do not have such a tool at their disposal.

For example, one of the most involved areas of astronomy is the observation of meteors and meteor showers. Here, the naked-eye observer actually has an advantage over an amateur using a pair of binoculars or a telescope, for much more of the night sky can be observed with the naked eye than it can

© Springer Nature Switzerland AG 2018

J. Powell, *Rare Astronomical Sights and Sounds*, The Patrick Moore
Practical Astronomy Series, https://doi.org/10.1007/978-3-319-97701-0_5

be when focusing in on one particular object. By close study of any shower, one can determine over a period of time whether or not the shower stream is as active as previous years, or whether there has been a shift to a more bountiful display, or, by the same token, a decrease in output. The study of the type of meteors being generated by a certain shower is also very useful.

Whether it be simply observing sporadic meteors on a typical night or fully documenting the activity of a shower, having eyes trained on all parts of the sky to watch for meteors is something that the amateur astronomer of any level can take part in, submitting reports and findings that contribute towards an overall picture of this field of astronomy.

The Telescope: Early Days

Where was the concept of the telescope born? Phoenicians cooking on sand first discovered glass around 3500 BCE, but it wasn't until 5,000 or so years later that glass was eventually shaped into a lens to create the first telescope.

The Phoenicians were notable merchants, traders, and colonizers of the Mediterranean in the first millennium BCE. Phoenicia was an ancient region corresponding to modern day Lebanon, with adjoining parts of modern Syria and Israel. The Phoenicians and indeed Canaanites were renowned sailors and were regarded as excellent astronomers by ancient writers such as Strabo and Aratus. If their highly skilled maritime navigation is any indication, it is quite correct to presume that their knowledge of the night sky was great even by modern standards.

While there remains much speculation as to who invented the telescope, German–Dutch spectacle maker Hanns Lippershey (1570–1619) is commonly associated with the discovery, as he was one of the first to try to obtain a patent for it. One story of the discovery says that two children were present in his optics shop one day, playing with lenses that, when combined, were making the image of a distant weather vane appear closer. Another story claims that Lippershey stole the design from fellow eyeglass maker Zacharias Janssen (1585–pre-1632).

Lippershey's patent, filed in 1608, is the earliest written record of a refracting telescope. He made the application to the States General of the Netherlands on October 2, 1608, with the accompanying documentation stating it was "for seeing things far away as if they were near." It appears Lippershey's patent was submitted just weeks before that of Dutch instrument maker Jacob Metius (1571–1624/1631) who before his death sadly destroyed evidence of his other inventions so that nobody could steal his ideas. Metius further commanded that in his will, all of his tools were to also be destroyed in order to prevent anyone else from claiming the honor of inventing the telescope.

Both Lippershey's patent and Metius' submission were discussed at the States General. Metius has proposed a convex and concave lens in a tube, the combination of which would magnify an object three or four times. To complement his documentation, Metius informed the States General that he was familiar with the secrets of glassmaking, and that with the support of the State General, he could make an even better telescope. There was some reservation over his claims, and the States General were reluctant to review it, to a point where Metius subsequently refused to allow anyone to see his telescope. Yet despite this resistance to Metius and his claims, the States General made a decision following a vote to offer a small award to Metius to construct binocular versions of the telescope. Ironically, Metius turned to Lippershey, whose patent had not been accepted, to undertake the work.

The muddied waters concerning who invented the telescope are further discolored by Johannes Zachariassen, son of Dutch spectacle maker Zacharias Janssen, who claimed that Metius and Cornelis Drebbel bought a telescope from him and his father in 1620 and copied it. There is much confusion over the dates, as this occurred at a much later date than it was known that Metius was making telescopes.

Still, claims that Zacharias Janssen invented the telescope and/or microscope in Middelburg (a Dutch city that played a seminal role in the scientific revolution) remain. At one time in this rather secretive and highly competitive world of optics, Janssen actually lived next door to Lippershey. At the time, Dutch diplomat Willem Boreel (1591–1668) was tasked with trying to ascertain once and for all who had the rightful claim. With the help of a local magistrate, Janssen's son testified that his father had indeed invented the telescope and the microscope, dating the discoveries back to 1590. Boreel was convinced by the testimony, concluding that Janssen invented the telescope a little ahead of Lippershey, a finding that was adopted by chemist, alchemist, physician, botanist, and writer Pierre Borel in his 1655 book on the subject, entitled *De vero telescopii inventore* ("Inventor of the telescope").

Some would still argue that Lippershey developed the telescope independently of any outside influence, and should therefore be attributed with the discovery, whilst Jansen should be credited with the invention of the compound microscope. Both appear to have contributed to the development of the two instruments in some way.

Italian astronomer Galileo Galilei (1564–1642), the son of a musician, significantly improved on the work of Lippershey, Metius, and Janssen. Galileo's initial spark of enthusiasm was first generated upon hearing of the existence of the spyglass in 1609. Following his new passion, Galileo began experimenting with telescope-making and grinding out and polishing lenses.

Galileo's improvements of the original workings of the optical aid produced a telescope that far exceeded the limited magnification of the spyglass. Whereas the spyglass could manage at best a magnification of three, Galileo's telescope allowed for a factor up to eight or nine times. With this superior optic, it was not long before he turned his telescope to the skies, studying the craters of the moon before making detailed tracking observations of the phases of Venus. With more caution, Galileo viewed the Sun, discovering sunspots. Most puzzling of all were the rings of Saturn, which for him did not appear as rings but as two globe-like features on either side of Saturn's disc. Galileo was making far more progress in astronomy than many of his contemporaries, even if explanations for some of what he had seen weren't readily forthcoming.

It was his discovery of the four inner moons of Jupiter—Io, Europa, Ganymede and Callisto—that left Galileo's greatest impact on the world of astronomy, which he describes in his book *Sidereus Nuncius* ("Starry Messenger"), published in 1610.

Galileo is thought to have also made the first recorded studies of the planet Neptune, though at the time he did not recognize the point of light as a planet. The possible reference to Neptune was documented while Galileo was observing Jupiter's moons in 1612 and 1613, spotting a nearby star whose position was not found in any modern star catalogues.

In 1615, it was declared heresy if anyone challenge the accepted system that celestial bodies orbited the stationary Earth, a directive supported by the Catholic Church. Galileo nevertheless did not agree, with his aforementioned research of the heavens revealing the Copernican system to be incorrect. His observations of the phases of Venus and the Jovian moon system that did not orbit the Earth directly conflicted with the Copernican model.

Galileo was summoned to Rome and warned not to teach or write about his controversial theory but, believing that if he wrote on the subject using mathematics as the basis for his proposal, this would make his argument acceptable, Galileo continued to pursue his beliefs. Upon publishing his work against the Copernican system, Galileo was summoned again in 1632 and duly charged with heresy, spending the remaining years of his life under house arrest.

Despite this controversy, the Galilean telescope had already been born and was destined to spread. At the time, the Venetian Senate in Venice was about to purchase one of the popular gadgets of the time, the spyglass upon which Galileo had made his improvements. However, Galileo, then mathematics professor at the University of Padua, stepped in with his own version, which was constructed of wood and leather, with a convex main lens and a concave eyepiece, just like the original. Arguing for its superior power compared to the spyglass, Galileo took a number of senators up a nearby bell tower to show them exactly what his telescope was capable of, observing ships out in the lagoon. The senators showed interest, especially as any investment in such an instrument would be commercial rather than scientific. With sea attacks by the Turks threatening Venice's much-treasured overseas trade, such a tool would be very useful indeed, given its power to spot a potential enemy strike on Venice's shipping.

In this famous story lies some dispute as well, with English astronomer and mathematician Thomas Harriot (1560–1621) claiming that months before Galileo published his observations, which included maps of the mountains and craters on the Moon, he had already done so! Four months previously on July 26, 1609, Harriot claimed to have made a drawing of the moon through a telescope. The sighting of Halley's Comet in 1607 initially turned Harriot's attention towards astronomy. Harriot, who had been working on the theory of color at the time, was corresponding with German astronomer Johannes Kepler when his attention was drawn to the comet. Although nothing transpired from his dialogue with Kepler with regard to his work on the theory of color, the spark in astronomy had been ignited, and Harriot switched courses. Harriot's observations of Halley's Comet (which was only identified as Halley's Comet on September 17, 1607) were later used by other astronomers in an attempt to compute the comet's orbit, such was the precision and detail of his data.

In 1609, Harriot purchased a "Dutch trunke" (telescope) that had been invented in 1608. He then turned his telescope to the skies, making a map of the Moon. Harriot also observed sunspots in December 1610, predating the work of Galileo. Between December 8, 1610, and January 18, 1613, Harriot made documented observations of sunspots on 199 occasions, allowing him to figure out the Sun's period of rotation. Harriot went on to develop lenses with greater magnification, with one capable of a magnification of up to 32 created by April 1611. On February 26, 1612, he also made observations of the already discovered moons of Jupiter.

Early telescopes such as Galileo's consisted of glass lenses mounted in a tube, which Sir Isaac Newton (1642–1727) discovered caused differing refraction when different colored lights passed through it. In order to solve the problem, Newton designed a telescope that used mirrors instead; this became known as the reflector telescope, in contrast to the design from Galileo's refracting scope.

Newton's idea for the reflecting telescope was not a new one, as Galileo and mathematician Giovanni Francesco Sagredo (1571–1620), who were close friends, had also discussed the possibility of using a mirror as the image-forming objective soon after realizing the concept of the refracting telescope. Reports also exist that Bolognese Cesare Caravaggi had constructed a reflecting telescope around 1626. Indeed, the idea that curved mirrors behave like lenses dates much further back in time, to at least Arab mathematician, astronomer, and physicist Hasan Ibn al-Haytham Alhazen (c.965–c.1040) in his work Kitab al-Manazir ("Book of Optics"), which had been widely disseminated in Latin translations by early modern Europe.

Other persons laying claim to the idea include astronomer and physicist Niccolo Zucchi (1586–1670), who is thought to possibly be the first person to have seen belts on Jupiter (May 17, 1630) and reported "spot"-like features on Mars in 1640. Zucchi's book *Optica philosophia experimentalis et ratione a fundamentis constituta* ("An optical philosophy and the experimental nature of the bases determined"), published in 1652–56, described in detail Zucchi's experiments conducted in 1616 using a curved mirror instead of a lens as a telescope objective, which could well credit him with the earliest known description of a reflecting telescope.

Zucchi's own interest in astronomy had been encouraged by Johannes Kepler, whom he met when he visited the court of Emperor Ferdinand II as part of a retinue on an emissary mission for the Pope. Zucchi, an official of the Jesuit house in Rome, received patronage from a number of different individuals, whom he dedicated various scientific work to. Though Kepler encouraged Zucchi in astronomy, his passion for the subject had long since existed. His own efforts, while considerable and of great scientific note for the time, were ultimately conducted with a rather poor, primitive version of the reflector. Zucchi's design did not provide a way to keep the head of the observer from intercepting most of the rays, which were needed to form the focal image. There are references in the book stating that Zucchi experimented with a concave bronze mirror in 1616, concluding that it did not produce a satisfactory image.

However, Zucchi's book was a triumph and a landmark publication, and it is thought to have influenced the work of both Newton and Scottish mathematician and astronomer James Gregory (1638–1675). Gregory described an early practical design for the reflecting telescope—the Gregorian telescope—in his 1663 *Optica Promota* ("Promotion of Optics"). In the book, Gregory pointed out that a reflecting telescope with a parabolic mirror would correct spherical aberration (an optical problem that occurs when all incoming light rays end up focusing at different points after passing through a spherical surface) as well as chromatic aberration (failure of a lens to focus all colors to the same convergence point), which was seen in refracting telescopes.

Gregory's design placed a secondary concave mirror with an elliptical surface past the focal point of the parabolic primary mirror, reflecting the image back through a hole in the primary mirror where it could be viewed. This revelation earned the scope the name of the Gregorian telescope. Ironically, while on paper all looked practical and workable, Gregory—having no skills in constructing such a telescope—had to find someone to build it!

The design attracted several people, one of whom was English natural philosopher and architect Robert Hooke (1635–1703), who eventually built several telescopes some ten years later. Another who showed a keen interest in Gregory's work was Scottish soldier, statesman, and natural philosopher Sir Robert Moray (1608/1609–1673), founding member of the Royal Society of London for Improving Natural Knowledge.

Although the design of the telescope is rarely used in modern-day astronomy, with other types of far more efficient reflecting telescopes taking precedence, the principles behind Gregorian optics are still used in radio astronomy. One of the best examples is the Arecibo telescope in Puerto Rico, which features a "Gregorian dome" that contributes via the use of Gregorian optics to the imaging process of the telescope.

Despite all that went before it and all that came after it, Sir Isaac Newton's telescope marked a lasting turning point in the history and development of the instrument. Not only did its construction create a new dawn in optical astronomy, but it also fit into Newton plans to prove that white light is composed of a spectrum of colors. Newton who also acknowledged that the main failing of the refractor was chromatic aberration, set out to prove that this fault could be corrected, with his theory on white light the very driving force behind his own ideas for a reflecting telescope. Indeed, this immensely practical design is the main type of reflecting telescope used by amateur astronomers in the present day.

Binoculars in Astronomy

Before venturing into the world of astronomy, one should not discount binoculars and the many merits they present in their own right against the telescope. Many amateur astronomers begin their route to the skies via the naked eye, then with the use of binoculars, then a telescope. This rite of passage is a tried and true formula that allows any keen observer just starting out to learn how to walk before running. Binoculars can serve as a brilliant introduction to observing, and also as a lifelong companion for the amateur astronomer.

Binoculars present the observer with the easiest and simplest way to view the night sky. Apart from the obvious plus of being easy to use with a minimum amount of fuss, binoculars offer a diversity that telescopes do not. Should interest in astronomy wane, they can be used for other terrestrial observing purposes, and whilst telescopes can be used likewise, the problem of portability is evident.

Unlike telescope observing through a sole eyepiece, binoculars present the brain with two sets of incoming data via our eyes. The eyes, which are remarkably adapted for sensing light and color and for reacting to light and dark, provide the information; your brain provides the result, unscrambling and reassembling the data in order to make sense of it. Binoculars thus function as an extension of the natural abilities of the eyes in a more direct way than telescopes with their single eyepieces. They work by using a prism to lengthen the light path between the objective lens (that is, the front lens that captures the light) and the eyepiece. Lengthening the prism can thus increase the magnification without increasing the length of the binoculars.

Binoculars: Types

There are two main types of binocular design: roof prism and Porro prism, the latter named after the Italian inventor of optical instruments, Ignazio Porro (1801–1875), who in 1854 patented the image-erecting prism system contained within binoculars.

Each type of binoculars differs in the way the prisms channel light through and into the eyes of the observer. Roof prism binoculars have an H-shaped design, where the eyepiece and binocular tubes are in a single line. Constructed to be more durable and waterproof than their counterparts, such binoculars command a generally higher price in the marketplace, due the more complicated manufacturing process for the internal angled prism. Roof prism binoculars also tend to be a good deal smaller than Porro prism ones, with their overall design more compact in nature.

Porro prism binoculars present themselves with the more traditional M-shaped design, where the eye and the lens are not in line. They have a wide hinge between the oculars, a feature allowing the observer to adjust them to the size of their face.

Porro prism binoculars tend not to be waterproof and are typically less durable than roof binoculars, the driver in the marketplace for Porro prism binoculars being that many are built for a price. That is not to say that the optics are of lesser quality: price obviously dictates the quality of many items, and binoculars are no exception. Porro prism binoculars usually have a higher quality image and less light loss, which makes the image clearer. Roof prism binoculars can absolutely have a comparable optic, and their compact design and complex prisms make the higher-priced roof prism binoculars a better investment for a quality binocular.

Depending on which binoculars are chosen, roof prism or Porro, the observer can see up to 50 times more stars than with the unaided eye. This is not due to magnification alone, but the phenomenon of perspective-narrowing driving a flow state—the simple act of gaining focus and accompanying clarity on what is being observed, which has a measurable effect.

Binoculars: Size and Power

Binoculars can be generally defined by three categories: small, medium, and large.

Small binoculars represent the standard handheld kind, which are highly portable and lightweight. These tend to be the kind previously used for general, terrestrial-bound observations, with general interest at some point then turning to the night sky.

Medium binoculars, while offering a more powerful magnification, are generally still handheld, but the length of time of usage can be foreshortened, as the weight becomes more noticeable depending on how long the observer chooses to watch the skies. Frequent breaks may prolong the session, but inevitably arm-ache will prevail. Medium binoculars are slightly more difficult to negotiate, so the observer must apply a tighter brace when using them.

Large binoculars will need a tripod mount or counter-weighted arm, as they are virtually impossible to use as a handheld optical aid. Stability is key, and the sheer weight alone will not allow the observer much relief to center in on any object, given the bulkier size. Larger binoculars naturally command larger lenses, which in turn offer a fantastic field of view along with powerful magnification.

Aside from their actual size, binoculars have their capabilities emblazoned near the eyepiece: 8 x 25, 12 x 60, 25 x 100, for example. The two numbers are separated by an "x," which can be read as "by," as in 8 by 25 or 12 by 60. The first number refers to the measure of power, meaning how much the binoculars will magnify a certain object by; the second number is the metric diameter of the big round glass lens at the front of the binoculars. A pair of 12 x 60 therefore means that that pair of binoculars will make an object appear 12 times larger than the unaided eye would see it, with an objective lens that measures 60 millimeters across.

Considering the desired size of magnification is important, as it will have a bearing on exactly what is capable of being observed and how fine the detail is, which may make some closer astronomical objects like the Moon appear larger than desired, if lunar observing is the primary goal. One should also consider that without the aid of a tripod, 8x, 10x, and 12x magnifications are generally easier to hold steady than those with a larger magnification.

The second number as we have seen refers to the objective lens diameter, the lens through which light enters the binoculars. The larger the second number, the brighter the image in the binoculars will appear (all other factors being equal). Logically, the larger the objective lens diameter, the larger and heavier the binoculars are likely to be. Generally speaking, binoculars that carry an objective lens diameter greater than 30 millimeters are classed as standard-size. Those with an objective lens diameter of less than 30 millimeters are classified as compact.

Aside from considering intended use for the binoculars and whether or not their weight alone will be an issue, size also merits some thought. For example, take the objective size and divide it by the power. The number reached is the diameter of each of those small circles the observer will be looking through. Those disk-shaped bright spots are known as "exit pupils," with higher powered binoculars yielding smaller exit pupils. For example, a 12 x 60 pair of binoculars produces five-millimeter exit pupils; 8 x 60 binoculars present 7.5-millimeter exit pupils. In theory and practice, exit pupils should approximately match the pupils of the eyes of the observer, and, as we know, eyes change with age.

Binoculars: Field of View

Most binoculars have a field of view (FOV) number or numbers printed on them, in some cases expressed as an angle. The field of view is exactly how wide an area is encompassed in the image generated by the binoculars. The expression of field of view is described in two ways, that of angular field and linear field of view. Angular field of view can be used to calculate the linear field of view by multiplying the angular field by 52.5. For example, if the angular field of a particular pair of binoculars is eight degrees, then the linear field at 1,000 yards will be 420 feet (8 x 52.5). Usually, a higher power of magnification results in a narrower field of view. A percentage of binoculars will also give the observer the apparent field of view (AFOV). This is the value of the real field of view multiplied by the magnification of the binoculars (apparent field of view = magnification x real field of view). This value is important because it is comparable even along binoculars of different magnification.

Angular field of view is the true angle seen through the optics and is usually measured in degrees, while the linear field of view is the width of the area seen. A larger number for either angular or linear field of view denotes that a larger area will be observed.

If a pair of binoculars gives a wide field of view, it basically means that more of the image that is being observed can be "fitted in." This is beneficial for locating objects in the night sky, especially if the object is moving.

Aside from wide-angle binoculars, the marketplace also features ultra or extra-wide binoculars, which have huge fields of view. In astronomy, though some may argue the case, there doesn't seem to be much point in their usage, since the objects being pursued will appear so diminished that one might as well use one's own unaided eyes!

Binoculars: Eye Relief

The downside of a wide field of view is that such binoculars tend to have reduced "eye relief," and to gain the wide angle usually means reducing the magnification of the binoculars. Thus, whereas finding an object will be easier, the finer details are lost due to the lack of magnification. Eye relief refers to what the distance from the rear of the eyepiece lends to the exit pupil or eye point. It is the distance the observer must be positioned behind the eyepiece in order to see an image that does not show a reduction in brightness or saturation toward the periphery of vision compared to the image center. The longer the focal image of the eyepiece, the greater the potential eye relief.

The impact of eye relief is of particular importance for the observer who relies on spectacles (but not the observer who wears contact lenses). The eye of the observer who is reliant on glasses is typically farther away from the eyepiece, which necessitates a longer eye relief in order to avoid the aforementioned problems of brightness reduction and image saturation. Binoculars with short eye relief can also be hard to use in instances where it is difficult to hold them steady.

Eye relief is of course as much a factor for binoculars as it is for telescopes, but whereas eyepieces are readily confined with most binoculars, the many eyepiece options available for telescopes give the astronomer more personalized flexibility in achieving the best image possible.

The Telescope in Astronomy

A whole new world awaits the amateur astronomer with a telescope, for here, detailed observations can be made, accompanied by a change of roles from passive observation to true involvement.

A vast array of choices awaits the amateur astronomer, perhaps overwhelmingly so. Establishing what objects the observer wishes to see, the observing conditions that the observer views the sky from, and where the telescope will be housed or stored are questions that will push the astronomer's decision in a particular direction.

Aperture

The most important aspect of the telescope is the aperture, the diameter of the main optical component, which can either be the lens or the mirror. The size of the telescope's aperture determines how much light it can capture. The more light that is captured, the more objects you can see in the night sky. More light also means greater clarity in the images observed.

The "bigger is better" policy might at first seem like the best way forward when selecting a telescope. While this is essentially correct because of its superior light-capturing abilities, the observer will have to deal with a more cumbersome piece of equipment—fine if it has permanent housing in a purpose-built observatory in a backyard, but not so clever if the observer wishes to pack the telescope up for the evening and move to another position miles away to sight an object that cannot be viewed elsewhere. The observer must seriously contemplate all such viewing requirements. While considering that aspect, it is worth considering magnification, but not to an overly concerning degree, as it is not too inherently tied to the choice of telescope.

The magnification of a telescope is determined by the eyepiece used. Changing magnification means literally swapping eyepieces over, switching from a lower to a higher magnification and vice versa. Technically, any telescope can have an infinite range of magnification. Remember thought that too much magnification will only result in a useless blurred image.

Refractor

Refracting telescopes account for a fair proportion of telescopes on the market and are probably the most commonly used. The refractor uses lenses instead of mirrors, and the eyepiece is located in the bottom of the instrument.

Advantages

1. Easy to use
2. Excellent for lunar observations; discerning with some ease Saturn's rings, Jupiter's cloud belts, and its four main satellites; binary stars; some of the brighter nebulae and galaxies
3. Virtually maintenance free, with rugged construction keeping the optics safe

Disadvantages

1. Generally small apertures about three to five inches, which mean that observations of more distant objects such as galaxies can be poorer
2. Heavier, longer, and bulkier than an equivalent reflector of equal aperture
3. Good-quality refractors cost more per inch of aperture than any other kind of telescope

Reflectors

Reflecting telescopes use a mirror instead of a lens, with the eyepiece at the side of the main tube. They also carry their own advantages and disadvantages.

Advantages

1. The potential for larger apertures yields excellent viewing of deep-sky objects including galaxies, nebulae, and star clusters
2. A distinct difference in optical quality with low irregularities delivering very bright, clear, and sharp images
3. Costs the least per inch of aperture compared to refractors and Cassegrain telescopes, since mirrors can be produced at less cost than lenses

Disadvantages

1. Whereas the tube in the refractor is sealed, protecting the optics, the reflector tube is open to the air, which means dust and small particles can gather on the optics
2. Require more in the way of maintenance and care than refractors

Cassegrain Telescopes

Cassegrain telescopes use a combination of mirrors and lenses, making them catadioptric in design. Two of the most popular designs are the Schmidt–Cassegrain and the Muksutov–Cassegrain. Muksutov–Cassegrain (also known as "Maks") are generally more expensive to make than the Schmidt–Cassegrain ("SCTs") because of their more steeply curved corrector lens. Both telescopes use a corrector lens at the front of the telescope, as well as spherical mirrors, which induce spherical aberration (blurring). The corrector lenses eliminate this aberration, allowing the scopes to be relatively inexpensive compared to the other all-reflective Cassegrain designs, which require expensive aspherical mirrors. SCTs use a flat Schmidt corrector, which has a very slight aspheric curve; Maks use a meniscus corrector, a highly curved spherical lens, to achieve the same effect.

Advantages

1. Of the three—refractor, reflector, and Cassegrain—the Cassegrain is the most versatile for both lunar and planetary observations, along with deep-space astronomy.
2. A first-rate telescope for astrophotography and use of Charge-Coupled device (CCD) imaging
3. Best near focus of any type of telescope
4. Compact and durable with a closed tube design

Disadvantages

1. Generally more expensive than reflectors of equal aperture
2. Its appearance may not be suited to everyone's taste

Focal Length

The focal length is the distance between the lens and the eyepiece, which for many telescopes is a guide to its overall length. A telescope with a focal length of 900 millimeters would be just less than one meter long. The f-number is the focal length divided by the aperture, informing the observer whether or not the telescope is short and fat or long and thin. A telescope with a small f-number describes a short and fat design, giving the observer bright images of low magnification. A telescope with a larger f-number describes a long and thin design, giving dimmer images at higher magnification, assuming that the aperture is the same in both telescopes. However, Cassegrain telescopes are different, as their optical systems provide a long focal length and high f-number in a short tube.

Mount

There are two basic types of telescope mount for the observer to consider. The Equatorial mount allows the observer to follow the rotation of the sky as the Earth turns. The telescope needs to be set up carefully, with the polar axis pointing to the celestial pole, near Polaris. On completion, this will allow the telescope to keep objects in the field of view as the Earth rotates. Equatorial mounts generally cost a little more than their counterpart, the Altazimuth mount.

In contrast, the Altazimuth mount has a much simpler design, allowing the observer to swing the telescope up and down and left to right. However, as the mount does not automatically follow the rotation of the sky as the Earth turns, the observer will have to compensate by making necessary manual adjustments. The Altazimuth mount is cheaper and lighter for the same degree of stability.

Capturing the Rarities of Space

Charge Coupled Devices (CCDs) were invented in the 1970's and originally found application as memory devices. Their light-sensitive properties were soon being used for imaging purposes, producing a major revolution in the world of astrophotography. CCD's did so by dramatically improving the light gathering power of telescopes by almost two orders of magnitude.

CCD's work by converting light into a pattern of electronic charge in a silicon chip. This pattern of charge is then converted into a video waveform, digitized, and stored as an image file on a computer.

While a digital single-lens reflex camera (DSLR) can yield pleasing results, more intricate objects can be better captured using astronomical CCD cameras. These cameras come in two different versions, monochrome and "one-shot" color.

The monochrome version is more flexible than the one-shot color and is operated with an additional filter wheel and either: clear, red, green, or blue filters for galaxy imaging, known as Luminance, Red, Green, and Blue filters (LRGB filters); or narrowband filters for specialized imaging of emission nebulae, such as the Orion or Horsehead nebulae.

The one-shot works in a similar way to a DSLR, employing a Bayer matrix of color filters. It is optimized for astronomical use by being thermo-electrically cooled, producing 16-bit Flexible Image Transport System (FITS) files for best-quality pictures. FITS is the most commonly used digital file format in astronomy.

CCD cameras used in astrophotography require sturdy mounts to cope with vibrations from wind and other sources. To take the necessary long exposures to capture fainter objects, astronomers use a technique known as auto-guiding, which keeps the telescope locked on the target, preventing it from drifting out of focus. It can take many minutes of exposure time to capture a dim object, with auto-guiding allowing the astronomer to attain the best possible image clarity and sharpness. Most auto-guiders use a second CCD chip to monitor deviations during imaging. Aside CCD sensors, there are Complementary Metal-Oxide-Semiconductor (CMOS) sensors, although for the use in astrophotography, CCD sensors are seen more favorably.

In the search to capture and photograph *Astronomical Rarities*, the use of CCD imaging takes this astronomy to a different level.

The advent of computerized telescopes and image processing software has allowed observation of deep-sky objects even from light-polluted skies, with the CCD's cutting through the light pollution with specialized filters.

Chapter 6

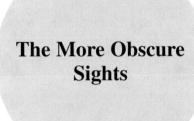

The More Obscure Sights

After learning to locate the many simple wonders of the night sky, from the craters on the Moon to Saturn's rings and beyond, the eagerness to observe rarities of an astronomical nature may entice the observer to seek out a more specialized field. For some, comets become the obsession, while for others, it is deep sky observations, with a proportion solely interested in asteroids and other small debris.

However, there are some sights that will elude the keenest of observers, possibly due to the obscure nature of the phenomenon or the time frame available to witness it. Some observers are fortunate to have seen some sights that others will never witness in their lifetimes, as with the transit of Mercury, but some might live in an age where a bright daylight comet will adorn the skies!

Mercury: Rare Transits

On average, Mercury transits occur just 13 or 14 times per century. One such transit took place in 2006, with one scheduled for 2019, then 2032. The first transit ever observed was of the planet Mercury in 1631 by French astronomer and mathematician Pierre Gassendi (1592–1655), earlier predicted by German astronomer and mathematician Joseph Kepler (1571–1630). A transit of Venus occurred just one month later, but Gassendi's attempts to observe it failed because the transit was not visible from Europe.

© Springer Nature Switzerland AG 2018
J. Powell, *Rare Astronomical Sights and Sounds*, The Patrick Moore
Practical Astronomy Series, https://doi.org/10.1007/978-3-319-97701-0_6

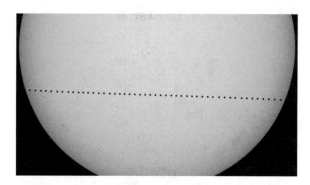

Fig. 1 Transit of Mercury, May 9, 2016. Courtesy of NASA's Goodard Spaceflight Center/SDO/Genna Duberstein

Transits are so rare because the innermost planet's orbit is inclined by about seven degrees compared to that of the Earth, so Mercury, the Sun, and the Earth only occasionally line up. Mercury completes one lap of the Sun every 88 days; the planet therefore crosses the plane of the Earth's orbit every 44 days, once while moving "up" and again while coming back "down." These particular points of interest are known as nodes. The nodes line up with the Sun from Earth's point of view just twice per year, once in May and again in November. If Mercury happens to be at the node at either of these times, Earth observers will witness a transit.

However, there are vast differences between the transits that occur in May and those that occur in November, with November transits only 10 arcseconds in diameter, as Mercury is near perihelion. An arcsecond, also called a second of an arc, is a unit of measurement that amounts to one sixtieth of an arcminute, equal to 1/3600 degrees of an arc. Used in other sciences, this unit of measurement is extremely important in astronomy, where the apparent size of a celestial object is often gauged by angular means. Most of the time, these angles are so small that they can only be denoted using arcminutes or arcseconds, as in the case of Mercury's 10 arcseconds during its November transit. By comparison, when Mercury is near aphelion, the disk appears 12 arcseconds across. Nevertheless, the probability of a May transit is smaller by a factor of almost two, as Mercury's slower orbital motion at aphelion makes it less likely to cross the node during the critical period.

Astronomers have a certain fascination with the Mercury transits, not just because of their rarity but also because close inspection of the phenomena can glean all sorts of insights about space. Back in 1677, Edmond Halley (1656–1742) observed Mercury as it crossed the Sun and noted that if a transit were seen from different latitudes on Earth, different observers would all see Mercury cross the Sun along a slightly different angle.

Those various angles could then be used to calculate the distance between the Earth and Sun, which at the time was still something of a mystery. Decades after Halley died, during the 1761 and 1769 transits of Venus, scientists from around the world collaborated and used Halley's method to calculate that the distance between Earth and the Sun was about 24,000 times Earth's radius, only about 3% off the real value!

During the November 15, 1999 transit, the centers of Mercury and the Sun were separated by 0.27 degrees. Because this degree of separation is so comparatively large, the centers were far away from one another and passed close to the edge of the Sun. Some parts of the world saw all of Mercury transit, while other parts only saw a fraction transit. The last time this happened was in the year 743, and will not happen again until 2391.

The transits in more modern times have been used to study the thin layer of gases that surround Mercury, known as an "exosphere." Sodium in the exosphere absorbs and re-emits a yellow-orange color from sunlight, and by measuring that absorption, it is possible for astronomers to learn about the density of the gas that surrounds Mercury.

In order to witness the event, the observer must possess a good telescope with a proper solar filter to see Mercury silhouetted against the Sun. This is because Mercury is incredibly small compared to the Sun, only 1/160 the diameter. For example, using an 80-mm refractor, a magnification of 67x is required to see Mercury's tiny disk.

The image is so small that projecting it using a simple pinhole camera (as many observers do to view solar eclipses) will not produce good results. Solar filters can be used on both binoculars and telescopes to observe the spectacle, but caution is advisable. Checking with the manufacturers is a must to ensure the right filters are used.

For the best possible results, incorporating the best level of safety when the Sun is involved, the image should be projected using a refractor or small Newtonian telescope. With a low-power eyepiece, simply align the telescope using the shadow of the Sun on the ground. Never attempt to align the telescope by looking through the eyepiece of the finderscope!

With a piece of white paper positioned around 30 centimeters away from the eyepiece, the image of the Sun's disc with the tiny disc of Mercury should be evident on the sheet of paper.

Venus: Even Rarer Transits

Venus also transits the Sun, but because Venus circles the Sun much more slowly than Mercury does (once every 225 days, compared to once every 88 days), Venus transits are much rarer, with only eight such events having occurred since the invention of the telescope in the early 1600s.

Another factor contributing to its rarity is that the planet's orbital plane is out of alignment with the Earth's by about three degrees.

In 1639, English astronomers Jerimiah Horrocks (1618–1641) and William Crabtree (1610–1644) became the first to witness a transit of Venus. The pair observed the transit from different parts of England—Horrocks from Much Hoole in Lancashire and Crabtree from Broughton, near Manchester. Along with another astronomer from the same region in England, William Gascoigne (1612–1644), they followed the work of the then-new astronomer Johannes Kepler (1571–1630). Horrocks correctly predicted the transit of Venus, which Kepler had failed to predict because of errors in his calculations.

Venus transits happen in pairs eight years apart, with each pair separated from the previous pair by more than 100 years. The sequence of the last eight occurred in 1631, 1639, 1761, 1769, 1874, 1882, 2004, and 2012. The next Venus transit is not due until 2117.

While *Astronomical Rarities* covers just that, barring any major medical breakthroughs, most of us won't see one at all, or in some cases, ever again!

Sunspots

Sunspots are darker, cooler areas on the surface of the Sun in a region called the photosphere. The photosphere has a temperature that ranges between 4,230 and 5,730 °C, with sunspots emitting temperatures of about 3,600 °C. The composition is made of two areas, a noticeably darker umbra region, with a lighter part surrounding the mass, the penumbra.

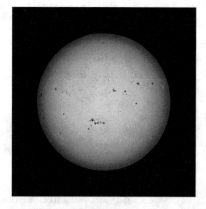 

Fig. 2 Sunspots. Left: July 19, 2000, around the solar maximum. Right: March 18, 2009, around the solar minimum. Courtesy of NASA/Solar and Heliospheric Observatory (SOHO)

Sunspots appear dark only in comparison with the brighter and hotter regions of the photosphere that surrounds them. If we could take a sunspot out of the Sun and place it in our night sky, it would be as bright as the full Moon. Sunspots can be very large in size, measuring up to 50,000 kilometers in diameter, although there can be great variations in size and shapes. Sunspots are caused by interactions with the Sun's magnetic field. They occur over regions of intense magnetic activity, and when the energy is released, solar flares and large storms called coronal mass ejections can erupt. Some groups of sunspots have a more complex magnetic structure than other sunspot groups and are more likely to produce solar flares.

Determining whether or not a particular sunspot group is a potential threat for a stronger-than-usual solar flare is not an easy task, given the variants in size, temperature, and location of the grouping on the Sun's surface. To tackle the problem, the Mount Wilson Observatory in California made rules to assign every sunspot region a certain magnetic classification. Every single day, space weather specialists count the number of sunspots, with every sunspot group receiving a number, a magnetic classification, and a spot classification.

The daily (synoptic) sunspot drawings done at the 150-Foot Solar Tower at Mount Wilson began on January 4, 1917. Previously, sunspot sketches were often made by observers, but it was not until a parallel-plate micrometer was built to visually measure sunspot magnetic field strengths that solar drawings became part of the regular observing routine. All of the drawings since 1917 have been archived. As of 2007, there are over 27,000 on file. The drawings are used by the Space Weather Prediction Center in Boulder, Colorado, to help forecast solar flares. The agency correlates findings from Mount Wilson with similar observations and data from Australia, Italy, and New Mexico. While in more recent history, telescopes and cameras on spacecraft have been able to generate more detail than the solar telescope at Mount Wilson, the human eye remains generally better at identifying and interpreting subtle differences and features of each sunspot.

Sunspot observation at Mount Wilson's Solar Tower is conducted by the adjustment of two mirrors that aim the sunlight through a lens down a narrow shaft to the bunker, reducing the diameter of the Sun from 865,000 miles to just 17 inches. The image then sits on a table within the observation room, and it is here that the detailed sketching of sunspots can begin, using a 10 × 20-inch paper, which is inserted to accommodate the disc of light.

The task requires much dedication, with one particular astrophysicist, Steve Padilla, having devoted many years of his life to the daily routine of producing pencil drawings of sunspots, supplying Mount Wilson with the longest continual record of sunspot activity with magnetic data ever made.

The different sunspot classifications are:

1. **Alpha:** A unipolar sunspot group
2. **Beta:** A sunspot group that has a positive and negative polarity (or bipolar) with a simple division between the polarities

3. **Gamma:** A complex region in which the positive and negative polarities are so irregularly distributed that they can't be classified as a bipolar sunspot group
4. **Beta-Gamma:** A bipolar sunspot group complex enough so that no line can be drawn between spots of opposite polarity
5. **Delta:** The umbrae of opposite polarity in a single penumbra
6. **Beta-Delta:** A sunspot group with a general Beta magnetic configuration, containing one (or more) Delta sunspots
7. **Beta-Gamma-Delta:** A sunspot group with a Beta-Gamma magnetic configuration, containing one (or more) delta sunspots
8. **Gamma-Delta:** A sunspot group with a Gamma magnetic configuration, containing one (or more) Delta sunspots

More than half of the observed sunspot groups receive an Alpha or a Beta classification. Bigger sunspots are often more complex and get a Beta, Beta-Gamma, or Beta-Gamma-Delta classification. It is well known that Delta sunspots can be very active and produce the most intensive solar flares.

Sunspots are a common sight on our Sun during the years around solar maximum. Solar maximum or solar max is a period of greatest solar activity in the solar cycle of the Sun, where one solar cycle lasts about 11 years. Around the solar minimum, very few or even no sunspots are seen.

The Sun rotates around its axis just as the Earth does, with features like sunspots therefore following its rotation. This means that a sunspot travels across the solar disk from east to west, as seen from Earth. This is important because sunspot regions need to be close to the central meridian (as seen from Earth) in order to send coronal mass ejections towards Earth.

It takes a sunspot region near the equator about two weeks to move from the eastern limb of the Sun to the western limb (as seen from Earth). The further away a sunspot region is from the equator, the longer it takes to move across the face of the Sun, as the Sun rotates faster at its equator than at its polar regions. This rotational period lasts 25.6 days at the equator and 33.5 days at the poles.

Largest Sunspots

One of the largest sunspots sighted on the Sun was captured by NASA's Solar Dynamic Observatory on October 23, 2014. The spot, named AR2192, measured nearly 129,000 km in diameter. Ten Earths could be laid across this massive length. However, the accolade for the largest sunspot on record goes to the Great Sunspot of 1947, which was 40 times the diameter of the Earth.

Sunspot Observation

As with the observing of Mercury transits, the preferred and safest method is to project an image of the Sun onto a piece of white paper. For more than just a quick look at the Sun, one might consider a projection screen attached to the telescope. Another option is to construct a projection box to improve contrast by keeping daylight off the paper.

Direct viewing is possible using specialized aperture filters. The most economical is made of Mylar plastic, which usually turns the Sun blue. The other type of filter is the metal-on-glass filter, which leaves the Sun with a more natural tint. While more durable, they do generally cost more. Be sure to keep the filters in perfect condition, as marks and scrapes can allow sunlight to penetrate through even the smallest of gaps.

Fig. 3 Projected images of the solar eclipse, August 21, 2017. Employees at NASA's Johnson Space Center in Houston experiencing a solar eclipse. (Courtesy of NASA)

Eclipses

From Earth we see two types of eclipses—eclipses of the Sun (solar eclipses) and eclipses of the Moon (lunar eclipses). There are several types of solar eclipses:

1. **Total Solar Eclipse:** This occurs when the Moon completely covers the Sun, as seen from Earth. Totality during such an eclipse can only be seen from a limited area, shaped like a narrow belt, usually about 160 km wide and 16,000 km long. Areas outside of this "track" may be able to see a partial eclipse of the Sun.
2. **Partial Solar Eclipse:** This occurs when the Moon only partially covers the disk of the Sun.
3. **Annular Solar Eclipse:** This occurs when the Moon appears smaller than the Sun as it passes centrally across the solar disk, and a bright ring, or annulus, of sunlight remains visible during the eclipse.
4. **Hybrid Solar Eclipse:** This is a rare form of solar eclipse, which changes from an annular to a total solar eclipse, and vice versa, along its path.

Types of Lunar Eclipses:

1. **Total Lunar Eclipse:** This occurs when the Earth's umbra—the central, dark part of the shadow cast by the Earth—obscures the Moon's entire surface.
2. **Partial Lunar Eclipse:** This type of eclipse is observed when only part of the Moon's surface is obscured by the Earth's umbra.
3. **Penumbral Lunar Eclipse:** This occurs when the Moon travels through the faint penumbral portion of the Earth's shadow.

A total solar eclipse is a very much sought after rarity. Solar eclipses occur about two to four times a year, but the area on the ground covered by totality means that not all parts of the globe can witness the event. In any given location on Earth, a total eclipse happens only once every hundred years or so, although for a few select locations, they can occur as little as a few years apart. An example of this is the August 21, 2017 eclipse and the April 8, 2024 eclipse, both of which were/will be visible from the same location (weather permitting)—Carbondale, in Illinois. Quite staggering, considering that on average, it takes about 375 years for a total solar eclipse to happen again at the same location.

The maximum number of solar eclipses that can take place in the same year is five, but this is rare. According to NASA calculations, only about 25 years in the past 5000 years have had five solar eclipses. The last time this happened was 1935; the next time will be in 2206.

A solar eclipse is quite magnificent—not just the actuality of the eclipse, but all the various other observable phenomena that are afforded during the

Fig. 4 Solar eclipse crown of white flare, August 21, 2017. The Sun's corona, only visible during a total eclipse, shows itself as a ring of white flares. (Courtesy of NASA)

experience. This was the case during the total solar eclipse on August 21, 2017. During this eclipse, the light faded dramatically, and the stars and planets on view at the time revealed themselves in a strange daytime glory.

About 15 seconds before the Moon completely covers the disk of the Sun, only a tiny crescent of sunshine is left, and the Sun's faint upper atmosphere, the corona, begins to come into view. Around this time, the sliver of bright sunlight transitions into a spectacular burst of radiance concentrated in one region along the Sun's edge, making the diamond ring effect. On either end of the two and half minutes of totality, a series of white glowing dots known as Baily's Beads, named after English astronomer Francis Baily (1774–1844), appear along the edge of the Moon's silhouette. This jewel-like display is formed by sunlight seeping through the deep valleys and craters along the edge of the lunar disk. Baily's name was linked to the phenomenon after his explanation of how they occurred, following a total solar eclipse on May 15, 1836.

During totality, the crown of the eclipse, a beautiful corona, appears in full view. Pearly white in color, the corona (the glowing upper part of the Sun's atmosphere) stretches out from the darkened Sun for millions of miles. Along with the corona, solar prominences can be observed—brightly colored tongues of hot hydrogen gas rising from the lowest layer of the Sun's atmosphere, the chromosphere. They reach out tens of thousands of miles into space, far enough to be visible above and beyond the eclipsing Moon as tiny flames of red and electric pink!

Lunar X

When the Moon appears 50% illuminated by sunlight (first quarter phase), the interplay of light and shadows provides a remarkable view of lunar craters large and small, which are easily visible in binoculars or a small telescope.

Fig. 5 Lunar X. Image captured of the Lunar "X" surrounded by craters. Courtesy of astrophotographer Miguel Claro at the Cumeada Observatory, headquarters of the Dark Sky Alqueva Reserve in Portugal

Among some of the more bizarre formations only visible during this time is the Lunar "X." When the terminator—the line between light and dark on the Moon—is located in just the right place, it seems as though the letter "X" appears on the Moon's surface. The illusion is created by sunlight falling on the rims and ridges between the craters La Caille, Blanchinus, and Purbach. To find the X, follow the terminator from the Moon's southern limb to about one-quarter of the way across the lunar disc. The phenomenon is only visible for a few hours every lunar month.

Various other letters can be found on the Moon, including the Lunar V shape in Mare Vaporum, sometimes prominent around the same time as the Lunar X. It may well be possible to see and photograph both at the same time. There is also a Lunar S in Sinus Asperitatis, visible at 47% illumination just before first quarter phase, along with a Lunar W located near Mons Rumker on the lunar limb in the Oceanus Procellarum. A Lunar Q also exists in Mare Nubium, reaching favorable illumination 10 days after New Moon.

Leonids

Although the Leonids meteor shower is not a rarity with its regular, annual appearance in the astronomical calendar, its ability to generate storm-like amounts of shooting stars makes it a notably exceptional sight for the astronomer.

Activity in the Leonids commences around November 14 and concludes a week later on November 21. This shower has the potential to deliver 100,000 meteors per hour. It is held in great esteem within the astronomical community, as the discovery and ensuing study of the stream marked the start of meteor astronomy. Observations of the Leonids stream laid down many of the fundamental principles for the scientific study of meteors. This revolution in thinking dispensed once and for all with the arguments that the phenomenon was merely one of an atmospheric nature and had nothing whatsoever to do with space (more on this shortly).

The first accounts of the Leonids shower dates back to 902 CE. In 898 CE, the yet-undiscovered comet that was to become the source of the debris for the stream crossed the path of Earth's orbit for the first time. Four years later, Chinese astronomers and observers in Egypt and Italy reported the first Leonid storm, an outburst that was intermittently recorded in the centuries that followed. On one occasion, sky watchers proclaimed that the "stars fell like rain."

In 1630, several days after the death of Johannes Kepler on November 15, 1630, the sky lit up with shooting stars from the then-unknown Leonids shower, at the time reckoned by some observers to be a salute from God to the German astronomer.

In 1799, German scientists and explorers Alexander von Humboldt (1769–1859) and companion Aimé Bonpland (1773–1858) observed an outburst from the Leonid shower in Cumana (latter-day Venezuela). Humboldt and Bonpland documented what they had seen in detail, releasing their findings to the scientific community.

On the night of November 12–13, 1833, and in particular the early morning hours of November 13, it was said that the skies literally filled with meteors, with some reports claiming that Judgment Day itself had arrived.

The Leonids were to put on another fine display in 1866, with observers documenting hourly rates as high as 5000, and a reduced rate of around 1000 in the following year. A similar Zenith Hourly Rate (ZHR), the rate at which a shower produces meteors, of 1000 was generated in 1868.

After Comet Tempel-Tuttle's closest approach to the Sun on January 12, 1866, data and subsequent calculations struck a firm link between heightened activity in the Leonids shower and that of the orbit of the comet. In 1867, Austrian astronomer Theodor von Oppolzer (1841–1886) calculated an orbital period for the comet of 33 years, meaning the next display of any note should occur before the turn of the century, in 1899.

Despite an initial good showing of the meteors in 1898, the Leonids did not live up to expectation. There was some cheer during the generally poor showing of 1899, with observers in 1901 documenting ZHR rates at 300–400 per hour from the western half of the United States.

The year 1932 also marked a decline in the Leonid rates and, despite some fairly lively displays up until the end of 1939 (30–40 per hour), the years now appeared to show that the shower's rates had reached a plateau. 1966 could only manage a near parallel to 1899, with around 100 meteors per hour being observed.

Despite the generally poor showing, the western half of United States was treated to a morning display of meteors not to be forgotten. In the pre-dawn skies of November 17, 1966, observers witnessed the rare spectacle of what was later estimated to be 40–50 meteors per second! Here in the skies of western North America it again "rained" meteors, with the tag of "Great Storm" being awarded to the shower by some fortunate observers. This event also revealed that the source of the outburst was in fact meteoroids released from the nucleus of Tempel-Tuttle some 66 years previously. Rates remained elevated for a time over the coming years, with a dip in 1970 and then a further upturn in 1971.

The 1990s produced more erratic displays from the Leonids shower, with the expected normal ZHR of around 10 to 15 an hour only reasonably heightened during 1994 and again in 1995, with 40 meteors per hour recorded. As with other years, the shower didn't fail to deliver a special showing for some parts of the world. In 1997, observers watching from the Canary Islands were treated to a ZHR of upwards of 2000.

In 1999, upon speculation of the Earth passing through a denser part of the debris left by Tempel-Tuttle, 1000 meteors per hour were predicted. The erratic nature of the shower far surpassed that estimate, with a ZHR in excess of 3000 meteors per hour.

In 2006, an outburst was predicted from the 1932 trail of debris left by the comet, with rates higher but not overly increased at a ZHR of 70–80. In 2007, the ZHR was higher than average but halved against 2006. The following year saw a rise in ZHR to over 100, this particular increase being associated with Earth passing through another denser area of debris. 2009 gave an impressive display to some observers across the globe, with a ZHR in excess of 500.

The possibility of increased rates with future displays remains, but even predicting when the Earth will pass through another significant clump of debris left by Tempel-Tuttle is not a guarantee, as other forces in the Solar System are at work to deprive observers of another big display! In 2028, Jupiter's gravitational influence is expected to seriously alter the orbital path of the comet, throwing Tempel-Tuttle off of its current path through space and making it all but impossible—at least through the beginning of the 22nd century—to see a repeat of the Great Leonid Storm of 1966. However, before Jupiter has significantly altered the path of the comet, some quarters point to potential outbursts of the Leonids in 2034 and 2035!

Halley's Comet

Being able to observe any comet can be quite a privilege, but there is one comet whose appearance fires the imagination more than any other: Halley's Comet. Named after Edmund Halley, who discovered it in 1531, Halley's Comet has been rounding the Sun every 75–76 years since, and for many years before that—sightings having been documented as far back as 240 BCE. A household name, Halley's Comet made its last appearance in 1986, which allowed for the first time the active involvement of spacecraft to make close-up observations.

Due to the comet's position relative to the Earth, the 1986 apparition was one of the worst viewing circumstances for seeing the comet. Its next return in 2061 is predicted to be a better one for observations, with perihelion occurring on July 28. The poor apparition of 1986 was due to Halley's Comet being at its brightest when it was closest to the Sun as viewed in our skies. Estimates for the 2061 return indicate a visual magnitude of −0.3, making for an easy naked-eye target. The comet will also be on the same side of the Sun as the Earth, allowing for much easier viewing. As for the future of the comet, barring a collision or a dramatic encounter with an object that alters its well-trodden path around the Sun, there is no reason why it should not continue its periodic route. It is likely that Halley's Comet has been on its current course for 16,000–200,000 years. Over time, the nucleus at the center has been significantly reduced, so perhaps the comet will eventually evaporate or split up within a few tens of thousands of years. Also, every orbit does alter slightly, meaning that over time, the course could be changed enough for Halley to simply leave the Solar System.

Planetary Alignment

The possibility of alignment between the planets in the Solar System is very rare. A recorded alignment of Mars, Saturn, Venus, Mercury, and Jupiter occurred in 2000. In May 2011 and again in 2015, it was also recorded that a triangular alignment of Jupiter, Mercury, and Venus occurred. Astronomers have calculated that there will be a rare planetary alignment of Mars, Mercury, Venus, Jupiter, Saturn, and the Moon in 2040.

Thunderclouds on Saturn

Every 30 Earth years, a huge storm formation occurs in the Northern Hemisphere of Saturn. This rare astronomical event is known as the "Great White Thunder Storm." This ammonia-rich cloud causes strong thunder and lightning, along with cloud formation spanning a length half the diameter of Earth.

It is estimated that there are ten string lightning occurrences every second within the atmosphere of Saturn. The lightning eventually vaporizes the planet's atmospheric water content. Once the atmosphere becomes more intense, thunderstorms develop, with this particular storm measuring up to 10,000 km wide, nearly as wide as the Earth, with a tail of white clouds that encircles Saturn.

These curious cloud formations on Saturn, which reveal themselves as white spots or white ovals, are observable from Earth, although the phenomena are usually short-lived, lasting from only a few days to a few months. Such a prominent white spot appeared in 1933, discovered by British comic actor and amateur astronomer Will Hay (1888–1949). Another spot was discovered in late 2010 by Australian amateur astronomer Anthony Wesley, subsequently becoming the most widely observed spot in Saturn's observational history. There is some scientific evidence to support the theory that such spots appear on Saturn's surface about every 57 years (that's the equivalent of under two Saturnian years).

Chapter 7

The Time, the Place

When and Where

One particular problem shared among the majority of observers around the world is a constant, irritating, and unresolvable aspect of astronomy—the weather! A changeable climate can be downright infuriating, with everything seemingly so close at the end of an eyepiece, but because of the weather, remaining so far away from observation. For many, their astronomical interest will be dictated by the elements, which is why, if an opportunity arises to escape a poor climate for a time, one might well consider this change of scenery in order to enjoy a much different perspective of the night sky.

The experience of a truly good, rich, and clear sky is a remarkable one. International Dark Sky Parks and Reserves provide a great starting point to seeing the skies as they were intended to be seen.

It is necessary to clearly define what designated areas serve what purpose. A Dark Sky Preserve (DSP) is an area usually surrounding a park or observatory that restricts artificial light pollution. The sole purpose of a Dark Sky Preserve is to promote astronomy. In 1993, Michigan became the first in the United States to designate a track of land as a Dark Sky Preserve, located at the Lake Hudson State Recreation Area. In 1999, the first permanent preserve was established at Torrance Barrens in the Muskoka region of Ontario, Canada.

© Springer Nature Switzerland AG 2018
J. Powell, *Rare Astronomical Sights and Sounds*, The Patrick Moore
Practical Astronomy Series, https://doi.org/10.1007/978-3-319-97701-0_7

Because different organizations across the world have worked independently to create their own programs, different terms have been adopted, in some cases meaning exactly the same thing. The International Dark-Sky Association (IDA) has chosen "reserve" to avoid confusion with "park" when using the acronyms "IDSR" (International Dark Sky Reserve) and "IDSP" (International Dark Sky Park). The IDA is a non-profit organization working to stop light pollution and protect the night skies for present and future generations.

Often, owners and administrators of public and private land with an "exceptional or distinguished" quality of starry skies unite and form a partnership, which then commits to protecting the visibility of the night skies with regulation, formal agreement, and long-term planning.

The IDA has five types of designations:

1. **International Dark Sky Communities:** Communities are legally organized cities and towns that adopt quality of lighting ordinances and undertake efforts to educate residents about the importance of dark skies.
2. **International Dark Sky Parks:** Parks are publicly or privately owned spaces protected for natural conservation that implement good outdoor lighting and provide dark sky programs for visitors.
3. **International Dark Sky Reserves:** Reserves that consist of a dark "core" zone surrounded by a populated periphery where policy controls are enacted to protect the darkness of the core.
4. **International Dark Sky Sanctuaries:** Sanctuaries are the most remote (and often darkest) places in the world whose conversation state is most fragile.
5. **Dark Sky Developments of Distinction:** Developments of Distinction recognize subdivisions, master planned communities, and unincorporated neighborhoods and townships, the planning actively of which promotes a more natural night sky. These do not qualify for the International Dark Sky Community designation.

A system of Gold, Silver, and Bronze Tier designations was instigated in order to reflect an ascending order of quality of observing conditions and the related air and light pollution.

There are further designations, including a "Dark Sky Nation," given to the Kaibab Indian Reservation. Located in the northern part of Arizona on the border with Utah, the land covers 488.9 km^2, inhabited by a federally recognized tribe of the Kaibab Band of Paiute Indians.

Another specific designation is the "Parashant International Night Sky Province–Window to the Cosmos," given to the Parashant National Monument located in northwest Arizona.

Alongside the designation work of the International Dark Sky Association, we must consider the Bortle Scale. The scale is a nine-level numeric scale that measures the night's brightness at a particular location. The scale quantifies the astronomical observability of celestial objects and the interference caused by light pollution. Created by American amateur astronomer John E. Bortle, it aims to help astronomers evaluate the darkness of an observing site and to compare the darkness of observing sites. The nine-level scale was born from Bortle's 50 years of observing the night sky and, as briefly defined below, marks the varying kinds of locations. From the very best observing conditions of a Class 1, which allows a great detail of many astronomical phenomena to be observed, to Class 9, where only the brightest constellations and stars are visible, this scale provides a reliable, quick guide to the different observing locations throughout the world:

Class 1–Excellent dark-sky site
Class 2–Typical truly dark site
Class 3–Rural sky
Class 4–Rural/suburban transition
Class 5–Suburban sky
Class 6–Bright suburban sky
Class 7–Suburban/urban transition
Class 8–City sky
Class 9–Inner-city sky

Globetrotting Astronomy

Different countries offer different astronomical phenomena, which, although in some cases can be sighted elsewhere in the world, are at their very best in a particular location. Armed with the knowledge of an upcoming astronomical event like a solar eclipse, astronomers can perfectly synchronize the time and place of observation.

The Northern Lights: Iceland, Finland, Norway and Sweden

The Aurora Borealis, or Northern Lights, are a spectacle that many have observed, but in order to really appreciate the true beauty and magnificence that such displays can offer, one has to travel to the so-called "auroral zone"—that offered by the chilling reach of the Arctic and all its glorious landscapes. The regions where the best opportunities await are at a latitude

Fig. 1 Aurora Borealis, November 3, 2015 in Talkeetna, Alaska

of 66 to 69 degrees north, a sliver of the world that includes northern Alaska and Canada, and portions of Greenland, northern Scandinavia, and northern Russia.

Here, the Northern Lights can be witnessed at their best, with the chances of catching a display significantly increased when compared to countries even only marginally to the south.

The Aurora Borealis is an incredible light show caused by collisions between electrically charged particles released from the Sun that enter the Earth's atmosphere and collide with gases such as oxygen and nitrogen. The lights are seen around the magnetic poles of the Northern and Southern Hemisphere. The southern lights are known as Aurora Australis.

Scientists have learned that in most instances, northern and southern auroras are mirror-like images that occur at the same time, with similar shapes and colors. Shapes can vary greatly in size, from small patches of light that appear out of nowhere to streamers and arcs that ripple across the sky in spectacular curtains, making for a tremendous show. Shooting rays of light have also been witnessed. The lights of the Aurora generally extend from 80 km to as high as 640 km above the Earth's surface.

Auroral displays can appear in many vivid colors, although green is the most common. Variations such as red, yellow, blue, and violet are seen on occasion. These changing colors are due to the type of gas particles colliding. Green, the most common color, is produced by oxygen molecules located about 95 km above the Earth. The rarer colors, such as red, are produced by high-altitude oxygen at heights of up to 320 km. The presence of nitrogen is responsible for turning auroras blue or purple.

Auroras not only occur on Earth but also on other worlds in our Solar System. Venus has an aurora generated by its stretched-out magnetic field known as a "magnetotail," Mars, which has too thin an atmosphere for global auroras, experiences local auroras due to the magnetic fields in the crust. NASA's MAVEN (Mars Atmosphere and Volatile Evolution) spacecraft also found widespread Northern Hemisphere auroras generated by energetic particles striking the Martian atmosphere. MAVEN was launched on November 18, 2013 from Cape Canaveral with a mission determine how the planet's atmosphere and water, presumed to have once been substantial, were lost over time. Data received from the MAVEN orbiter was published in 2015, confirming that the solar wind was responsible for stripping away the atmosphere of Mars over the years. Mars lost water into its thinning atmosphere via evaporation. In September 2017, NASA reported radiation levels on the surface of the planet that were temporarily doubled, which were associated with an aurora 25 times brighter than any observed earlier. This fluctuation was due to a massive unexpected solar storm in the middle of the month.

Jupiter, Saturn, Uranus, and Neptune all have thick atmospheres and strong magnetic fields, and each have auroras, although these auroras are a little different because they form under slightly different conditions.

The sunspots and solar storms that cause the most magnificent displays of northern lights occur roughly every 11 years as part of the solar cycle, which sees the Sun's magnetic field completely flip, meaning that the Sun's north and south poles switch places. It takes around 11 years for the Sun's north and south poles to flip back again. Solar Maximum and Solar Minimum refer respectively to the periods of maximum and minimum sunspot counts. Cycles span from one minimum to the next.

Does the Solar Minimum turn auroras pink? This was a phenomenon observed by sky watchers in the Arctic Circle in mid to late November, 2017. For the second winter in succession, unusual pink lights exploded in the night sky when the Sun was blank, showing no sunspot activity. The solar wind from the Sun's period of "low activity" produced auroras that were so bright they lit up the Arctic landscape, turning it pink.

In late October 2017, strange blue auroras were sighted, appearing as odd bands and rings over parts of Scandinavia and Russia. On this occasion, with no geomagnetic storm in progress, the answer had to be nearer to home. The explanation wasn't celestial at all: missiles fired by Russian forces during a war game exercise are thought to be responsible for the glowing blue cloud of rocket exhaust fumes around the Arctic Circle.

The Southern Lights: Antarctica, Argentina, and Remote Islands

The Aurora Australis, the Southern Lights, are best observed from Antarctica, but given that it is one of the most inhospitable places on the planet, visiting is more than a challenge. If the explorer in the observer shines through, and if enduring –50 °C temperatures, howling gales, and dangerous ice packs are of little challenge, then Antarctica will not fail to deliver the most visible, most active, and most impressive display of the Aurora Australis on Earth! Alternatively, the Southern Lights can be best seen from the southern tips of South America, Australia, New Zealand, and South Africa.

The world's most southerly city, Ushuaia, in Argentina, offers one of the best places to observe the Southern Lights, with March to September during the Antarctic winter being the peak time to catch any display. Despite Ushuaia's reputation for poor weather, the location still provides a very good location to see the lights.

A little more remote is South Georgia in the South Atlantic Ocean. A part of the South Georgia and the South Sandwich Islands British Overseas Territory, the area allows spectacular views of the phenomena.

The Stewart Islands off the southern tip of New Zealand also afford excellent opportunities. The islands have a large national park called Rakiura National Park. "Rakiura" is a Maori word that translates to "the land of the glowing skies."

The remote Falkland Islands, positioned 400 miles off the coast of Argentina, provide good views of the auroras between April and August, the darkest months on the island. In 2010, engineers on the island installed a monitoring system that was set to record activity of the Aurora Australis.

Germany

Designated in February 2014, the Westhavelland Nature Reserve is the closest International Dark Sky Reserve to a major populated city, situated approximately 100 km west of Berlin. Known in Germany as "Sternenpark

Westhavelland" (Star Park Westhavelland), it offers glorious sights of the Milky Way, along with rare displays of the Aurora Borealis, views of the zodiacal light, and possibly the gegenschein, a patch of faint light that can appear opposite the Sun in the night sky.

Aside from astronomy, the park offers much in the way of daytime activities, hosting thousands of migrating birds during autumn that by night make for a moving background noise under the stars. The park offers the best potential observing conditions between mid-May and mid-July.

The zodiacal light, or false dusk, is a light extending up from the western horizon. One may well have seen the phenomenon without realizing it. Originally thought to have somehow originated from phenomena in the Earth's upper atmosphere, the zodiacal light is in fact sunlight reflecting off dust grains that circle the Sun in the inner Solar System. These grains are believed to be the remnants left over from the process that created the Earth and other planets in the Solar System, dating back 4.5 billion years. Contributing to the substance of the grains are dust particles ejected by comets and fragments from asteroid breakups. This means that the debris is being constantly re-energized as both periodic comets and passing comets eject matter into the stream. This works the same way as a periodic comet that contributes additional material to debris it has left behind in previous visits, fueling a meteor shower.

The grains in space spread out from the Sun in the same flat disk of space inhabited by the Earth and the other planets. This flat disk, a narrow pathway that we refer to as the ecliptic, is the same pathway traveled by the Sun and the Moon across our sky, or as our ancestors first called it, the Zodiac or Pathway of Animals, in honor of the constellations seen beyond it.

Ranging in size from micro-sized to meter-sized, these grains are densest around the immediate vicinity of the Sun and extend outward beyond the orbit of Mars. Sunlight catches these grains to form the zodiacal light, and since they lie in the flat plane of space around the Sun, we could in theory see them as a band of dust across our entire sky, marking the same path that the Sun follows during the day.

Like the zodiacal light, the gegenschein (pronounced gay-gen-shine) is a phenomenon generated as sunlight reflects off these grains of interplanetary dust. From the German word for "counter-shine," the gegenschein is a faint, diffuse brightening along the ecliptic directly opposite and counter to the Sun. Appearing as a round to oval patch of light, this counterglow appears within the zodiac constellation crossing the southern meridian at that time. The glow is easiest to spot around midnight, when it will have reached its highest point.

Since the gegenschein lies opposite to the Sun, sunlight strikes the dust particles square on, with all the miniscule shadows hidden behind each and every grain, so much so that they don't subtract from the zodiacal belt's brightness, thereby creating a brighter spot in the sky. Our Moon experiences the same effect, with a similar heightening in brightness at the full Moon phase. It's not at all easy to sight, but under good conditions such as those offered the Westhavelland Nature Reserve, and in the absence of the Moon, it is certainly achievable.

Australia

Warrumbungle National Park in central western New South Wales was declared Australia's first Dark-Sky Park in 2016. Also the first Dark-Sky Park to be declared in the Southern Hemisphere, the location is a key area for observing the skies in Australia. Siding Spring Observatory is located nearby, housing the largest optical (visible light) telescope in the country. As Siding Spring is a working research laboratory, there are no public astronomy facilities. The park incorporates 24 acres of publicly accessible land with exceptional sky quality. Excellent and uncompromised views of the southern constellations await the observer.

Ireland

The Kerry International Dark Sky Reserve encompasses 700 km^2 on the Iveragh Peninsula, the location situated between the Kerry Mountains and the Atlantic Ocean, a natural oasis shielded from the light pollution of neighboring Irish cities. The Iveragh Peninsula, or "Ring of Kerry," is the largest peninsula in southwestern Ireland, the "ring" referring to the 179-km road that encircles this corner of the country. Situated south of the Dingle Peninsula and north of the Beara Peninsula, the area also boasts the "Kerry Way," the longest walking route in the Republic of Ireland.

The reserve offers excellent views of the Milky Way, star clusters, and nebulae, and it is perfect for observing meteors. Designated Ireland's first International Park by the International Dark Sky Association in January 2014, Kerry is also one of a select band of Gold Tier Reserves on the planet, meaning that the skies have been deemed absolutely stunning, akin to the observing conditions offered by the Grand Canyon or the plains of Africa. Other places in the world with the Gold Tier status include NamibRand Nature Reserve in Namibia, and Aoraki Mackenzie in New Zealand.

France

The Pic du Midi International Dark Sky Reserve attracts those wishing to gaze at constellations, the Milky Way, and perhaps sight the Zodiacal Light. Designated in December 2013, Pic du Midi covers 3,112 km^2 spread across the Pyrenees National Park and the Pyrenees–Mont Perdu, a UNESCO World Heritage Site.

The Park contains the Observatoire Midi-Pyrénées, situated at an elevation of 2,877 meters on the Pic du Midi in the Pyrenees. The observatory houses a two-meter Cassegrain telescope, named after French astronomer Bernard Lyot (1897–1952). The observatory has made several contributions to astronomy, including the observation of sandstorms on Mars and the conclusion that lunar soil behaves like volcanic dust.

It was Lyot himself who invented the coronagraph in 1930, a device that blocks light from the center of the telescope beam while permitting light from surrounding sources to pass through relatively undisturbed. Lyot's invention allowed astronomers to observe the hot gas (the corona) surrounding the Sun without having to wait for total solar eclipses. In 1939, using his coronagraph and filters, Lyot shot the first motion pictures of the solar prominence. His work in the field spawned the tribute Lyot Project, which used a stellar coronagraph to image faint objects surrounding other stars, rather than the coronae. The technique is essentially the same.

The Observatoire Midi-Pyrénées also houses one of the world's highest museums, with exhibits on astronomy and the history of Pic du Midi.

United Kingdom

Exmoor National Park is an area of hilly open moorland in West Somerset and North Devon in South West England, United Kingdom. The park was Europe's first International Dark Sky Reserve when it was designated in October 2011. On the clearest of nights, some 3,000 stars are visible. The United Kingdom offers the darkest skies in March and April, making for particularly good months to observe. Wimbleball Lake provides one of the best sights of the night sky. It has unobstructed lakeside views, and its distance from residential areas leaves the sky above in pristine condition.

The Brecon Beacons National Park in Powys, Wales, United Kingdom, was designated an International Dark Sky Reserve in February 2013, with the intention of preserving night skies, reducing energy waste, and preserving nocturnal wildlife. At that time, it became Wales' first and only the fifth destination in the world to be granted the status, joining Mont Megantic in

Quebec, Canada, and Exmoor National Park in South West England, among others.

Aside from the Brecon Beacons, Wales also has the Snowdonia National Park, which is an International Dark Sky Reserve; the Elan Valley, an International Dark Sky Park; and the Pembrokeshire Coast National Park, a Dark Sky Discovery Site.

Chile

The Atacama Desert in Chile stretches for 600 miles along the northern flank of the country, with high altitude, unpolluted skies, and the driest (non-polar) air on Earth. The ALMA Observatory, which contains the world's most powerful collective bank of radio telescopes forming the Atacama Large Millimeter Array, is based here.

Namibia

The NamibRand Nature Reserve resides on the eastern edge of the Namib Desert. Its low humidity sets excellent conditions for potential viewing of the night sky. With the nearest town 100 km away, the sky offers views of the Magellanic Clouds, the zodiacal light, and gegenschein. This private reserve in southwestern Namibia covers 2,150 km², sharing a 100-km border with the Namib-Naukluft National Park to the west and the Nubib mountains to the east. In 2012, the reserve gained the status of International Dark-Sky Reserve, an acknowledgement of some of the darkest skies yet measured.

With its red sand dunes and white salt pans, the terrain is almost as alien as the Magellanic Clouds above. These clouds are two irregular dwarf galaxies only visible from the Southern Celestial Hemisphere and are members of the Local Group orbiting our own Milky Way galaxy. These companion galaxies are named after the Portuguese navigator Ferdinand Magellan (1480–1521), whose crew discovered them during the first voyage around the world, 1519 to 1522. American astronomer Edwin Hubble (1889–1953) established that the clouds were of an extragalactic nature, separate systems from the Milky Way.

The clouds themselves are irregular galaxies that share a gaseous envelope and lie about 22 degrees apart in the southern night sky. The Large Magellanic Cloud (LMC) is a luminous patch measuring about five degrees in diameter, with the Small Magellanic Cloud (SMC) measuring less than two degrees across. Both visible to the unaided eye, the LMC is about 160,000 light years from Earth, measuring 14,000 ly in diameter, while the SMC is approximately 190,000 ly away, measuring 7,000 ly in diameter.

The LMC is classified in galaxy terminology as a Disrupted Barred Spiral, containing around 30 billion stars, with the SMC classed as a dwarf galaxy with three billion stars. There are three main types of galaxies: Elliptical, Spiral, and Irregular; two of these three types are further divided and classified into a system known as the tuning fork diagram because of the shape in which the diagram is traditionally represented. The system was devised by Edwin Hubble in 1926.

Research suggests that at one stage, the SMC could have been just one singular galaxy, with gravitational influence from the LMC tearing the single galaxy into two separate entities, although the whereabouts of the other segment are in doubt. The LMC is the third closest galaxy to the Milky Way, the other two being a dwarf galaxy in Sagittarius and another dwarf galaxy in Canis Major, neither one visible to the naked eye.

The formation of LMC and SMC is estimated to be around the same time that our own galaxy came into existence, some 13 billion years ago. The Milky Way galaxy—which measures considerably more in diameter, about 140,000 ly across—"captured" both the LMC and SMC, presently placing them in orbit around our galaxy. The orbit has not been a steady one, with several tidal influences occurring between all three.

Both the LMC and the SMC contain numerous young stars and star clusters, as well as some much older stars. One of the star clusters contains R136al, the most massive star known, with a mass 265 times that of our Sun.

The clouds also provide a perfect study of stellar birth and evolution, in particular that of the Tarantula Nebula (also called 30 Doradus), an immense ionized-hydrogen region that contains many young, hot stars. The total mass of 30 Doradus is about one million solar masses. With a diameter of 550 ly, it is the largest region of ionized gas in the entire Local Group of galaxies.

New Zealand

Situated on the South Island of New Zealand is the Aoraki Mackenzie International Dark Sky Reserve. Privy to some of the best sights of the Southern Hemisphere, including Aurora Australis, the Southern Cross, and the Southern Star, the reserve is one of the largest designated by the IDA, measuring almost 4,500 km^2.

Designated in 2012, the reserve is home to the University of Canterbury's Mount John Observatory, New Zealand's premier research observatory. Situated 1,029 meters atop Mount John at the northern edge of the Mackenzie Basin, the site houses many telescopes, including a 0.4-meter, two 0.6-meter, one 1-meter, and a 1.8-meter scope.

The largest telescope, not just at the facility but in the whole of New Zealand, is the 1.8-meter prime focus reflector, which began service in 2004. It was built by Japanese astronomers who dedicated the telescope to the MOA (Microlensing Observations in Astrophysics) project, an international partnership between astronomers at the Universities of Canterbury, Auckland, and Victoria in New Zealand, and Nagoya University in Japan.

It was reported in June 2008 that, after using the observatory's new MOA-II telescope, the smallest planet known at that time outside of the Solar System had been discovered. The planet, MOA-2007-BLG-192Lb, is just 3.3 times larger than the Earth, orbiting a small brown dwarf or low-mass star, MOA-2007-BLG-192L, 3,000 ly distant from Earth.

The planet was probably formed with lots of ice and gases, more akin to Neptune in composition than Earth. It has been speculated that the planet may well have a thick atmosphere and a liquid ocean on its surface.

United States of America

The southwestern US states of Arizona and New Mexico, along with California on the west coast, boast some of the best conditions for observing the skies, with a high density across the region of International Dark Sky Places, recognized by the International Dark Sky Association.

The Blue Ridge Observatory and Dark Sky Park in North Carolina are situated six miles west of Spruce Pine. The six-acre site is owned by Yancey County and managed by Maryland Community College. Designated in February 2014, the site hosts a public observatory and planetarium, featuring a custom-built 86-cm f/3.6 Newtonian telescope and a Meade 35-cm LX200 telescope.

Death Valley National Park, designated in February 2013, is the largest Dark Sky area with a span of 13,700 km. Despite the close proximity of Las Vegas (190 km to the southeast) and Los Angeles (460 km to the southwest) the park offers near pristine observing conditions.

The honor of the southernmost International Dark Sky Park, designated in February 2012, goes to Big Bend National Park in Texas. Its southern reach allows views of a portion of Crux, the Southern Cross and the Southern Stars. On a clear night, around 2,000 stars are visible to the naked eye. With an elevation of over 4,000 feet, the Park has a Gold Tier Dark Sky ranking. The park has the least light pollution of any other National Park unit in the lower 48 states. The dark skies are chiefly attributed to the remoteness of the area and the low population—a community of under 500 full-time residents. The site is located in an ancient sea bed known as the Magnificent Marathon Basin, surrounded by trees. Marathon measures Class 1 on the Bortle Scale, meaning that the sky is as dark as can possibly be achieved without actually going into space.

One of the country's largest public telescopes is housed in the Goldendale Observatory State Park in Washington, opened in 1973 with its designation bestowed in June 2010. The state park occupies five acres, with the primary instrument, a 24.5-inch reflector, housed in the South Dome. A Hydrogen-Alpha solar telescope is also in residence.

Mauna Kea ("White Mountain") in Hawaii is a dormant volcano on the island of Hawaii, the largest and southernmost of the Hawaiian Islands. Its summit places it above 40% of the Earth's atmosphere, providing conditions that allow observations of the faintest galaxies at the edge of the observable Universe. The air is thin at the top of the 4,205-meter summit, but despite the low oxygen levels, the prospect of witnessing some of the most spectacular sunrises and sunsets on Earth awaits. The site is home to the world's largest observatory for optical, infrared, and submillimeter astronomy. The combined light-gathering power of the telescopes on Mauna Kea is 15 times greater than that of the Palomar telescope in California and six times greater than that of the Hubble Space Telescope.

Canada

Designated with Silver Tier Status in September 2007, Mont-Mégantic International Dark Sky Reserve in Quebec was the world's first International Dark Sky Reserve. It is home to the largest telescope in eastern North America, which has an impressive mirror measuring 1.6 meters. This telescope is the fourth largest in Canada after those of the David Dunlap Observatory in Toronto, the Dominion Astrophysical Observatory in Victoria, and the Rothney Astrophysical Observatory in Alberta. Here, the Milky Way presents stunning views of itself under clear skies, with the Aurora Borealis also witnessed from this location.

On March 26, 2011, the Royal Canadian Astronomical Society (RASC) officially designated Jasper National Park a Dark Sky Preserve. Situated in the Canadian Rockies in the province of Alberta, north of Banff National Park and west of Edmonton, the park spans 11,000 km². Notable viewing locations in the park include Pyramid Island, Maligne Lake, Old Fort Point, and the toe of the Athabasca Glacier.

New Mexico

At the Chaco Culture National Historical Park, designated in August 2013, observers can view the heavens while standing amongst the ancient Pueblo ruins. Here, the Chacoan people observed the same night sky almost 1,000 years ago. The regulations of the Park strictly state no permanent outdoor lighting, with more than 99% of the area classed as a "natural darkness zone."

Designated in June 2010, the Clayton Lake Dark Sky Park in New Mexico offers some of the darkest skies in the country, with a computerized telescope and remote TV monitor providing live images of the night sky.

The Future Time and Place

Regardless of location, timing is critical to making the best of what the area has to offer, with one of the biggest enticements being that of a total solar eclipse. The arrival of a bright comet can also be a motivation to carefully plan out the best possible observing conditions.

In the decades to come, a revolution over accessibility of astronomical sights may well develop. The thrust of tourism seems less geared towards observing rarities and more focused on experiencing what it would be like to dwell within them, a prospect achievable through low-orbit spaceflight.

Reaching further, the Moon still poses the greatest challenge as a time and place to visit. In the future, there may well be an optimum time to achieve this as the priority switches from establishing a staging post on the Moon with the view to travel to Mars to the celestial body being bypassed altogether on a direct route to the red planet. This would pave the way for the Moon to become more of a tourist-focused destination. However, at present, leisure lunar travel remains firmly entrenched in science fiction, which is perhaps where it should stay.

Chapter 8

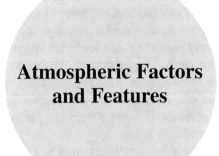

Atmospheric Factors and Features

Envelope of Protection

The Earth's atmosphere shields the planet from much but not all potential harm. In addition to being an abundant source of oxygen—a crucial resource for most life forms on Earth—the atmosphere also protects the planet's inhabitants from the intense light and harmful radiation of the Sun.

Without the atmosphere, which in essence forms the backbone of almost every biochemical process we know, humans would not exist. The composition of the atmosphere allows for the balance of life on Earth to be maintained, with the amount of carbon dioxide regulated through a balance between processes such as photosynthesis, respiration, and combustion—a balance that if destabilized could have dramatic consequences.

Imagine if our planet had no atmosphere. For an everyday comparison, look no further than our own Moon to view a body without an atmosphere. Suddenly, our planet would be just as silent, apart from the extreme surface temperatures and constant meteor strikes. Sound requires a medium to transmit waves. Vibrations could be felt from the ground, but nothing could be heard. Alongside the deafening silence, the sky, now void of birds and aircraft (because the air has no mass to support flying objects), would appear black overhead, because there would be no particles for the sunlight to interact with and scatter familiar blue light.

© Springer Nature Switzerland AG 2018

J. Powell, *Rare Astronomical Sights and Sounds*, The Patrick Moore
Practical Astronomy Series, https://doi.org/10.1007/978-3-319-97701-0_8

All plants and animal life would cease to exist, and with humankind unable to maintain life in a vacuum, all of those not in some way protected would also die. But death for humankind wouldn't be instantaneous. It would be likened to being drawn out of an airlock in space, as so many movies have depicted. If our atmosphere were to suddenly vanish, our eardrums would pop and our saliva would boil. Retention or attempting to hold one's breath would only result in one's lungs also popping, with an exhalation of any breath allowing 15 seconds of life before passing out, death ensuing around three minutes later. An oxygen mask would not alleviate the problem, as it would still be impossible to breathe. This is because your diaphragm uses the pressure difference between air inside your lungs and outside your body in order to inhale.

Without an atmosphere, rivers, lakes, and oceans would boil off, as whenever the vapor pressure of a liquid exceeds external pressure, boiling occurs. In the vacuum created by the lack of an atmosphere, water readily boils even if the temperature is warm. The resulting water vapor would not fully replenish the atmospheric pressure. An equilibrium point would be reached, where there would be enough water vapor to prevent the oceans from boiling off as the remaining water freezes.

With the vast majority of life wiped out, solar radiation would break atmospheric water into oxygen, which would react with carbon on the Earth to form carbon dioxide, still far too thin to be breathable. The lack of an atmosphere would soon chill the Earth's surface, whereby temperatures would drop below freezing, in turn causing water vapor from the oceans to act as a greenhouse gas (greenhouse gases being water vapor, carbon dioxide, ozone, and methane), subsequently raising the temperature. The increase in temperature would allow more water to transition from the sea into the air, with the likely outcome a greenhouse effect that makes the Earth more akin to our sister planet Venus, which we know to have a runaway climate.

Initially, anyone in a pressure suit and air would survive the calamity but, without an atmosphere to screen out solar radiation, the suit wearer would be charred inside the suit with extreme sunburn on their exposed skin. Away from the Sun, a similar extreme of cold, black emptiness would follow.

The construction of dome-like radiation-shielded housing would create a pressurized atmosphere in which humankind could exist, and with the environment able to support certain plants, life on a very basic level could slowly evolve. Are these the same prospects we face with possible human settlements on Mars? Despite the global devastation wreaked by the loss of an atmosphere, we can take a morsel of consolation knowing that some bacteria would survive it all, so life on Earth would continue nevertheless!

Life Without Oxygen?

Archaea are among the earliest forms of life that appeared on Earth billions of years ago. The archaea and bacteria developed separately from a common ancestor nearly four billion years ago, and millions of years later, the ancestors of today's eukaryotes split off from the archaea. Surprisingly, archaea are more closely related to humankind than they are to bacteria.

Archaea, which were initially classified as bacteria (receiving the name archaebacteria), convert a variety of substances for energy, including hydrogen gas, carbon dioxide, and sulfur. Archaea are microbes. Most of them live in extreme environments, earning them the name "extremophiles." Other archaea species are not extremophiles and live in ordinary temperatures and salinities.

Some extremophile species live in boiling water, like the geysers of Yellowstone Park and inside volcanoes. Their liking of the extreme heat has won them the nickname "thermophiles," which means "heat loving." The thermophile would most probably freeze to death at standard room temperature. Other extremophiles can be found in very salty water, called hypersaline environments. These salt-loving archaea are called halophiles.

In 2010, scientists discovered such creatures living in one of the harshest environments on Earth, the Mediterranean Ocean's L'Atalante basin, which contains salt brine so dense that it does not mix with the oxygen-containing waters above. Deep in the basin, about 3.5 km beneath the surface and about 200 km off the coast of Crete, Italian and Danish researchers described three new species of tiny animals called Loriciferans.

The Loriciferans live and reproduce in the sediments under the salt brine. The creatures' cells lack mitochondria, the organelles that use oxygen to power a cell. Instead, Loriciferans are rich in what seems to be hydrogenosomes, organelles that can do a similar job in anaerobic (oxygen-free) environments. The finding may help establish just what life looked like in the Earth's early oceans, which also had very little oxygen. Here, in the mud at the bottom of the L'Atalante basin, residing in oxygen-free water that has been in place for 50,000 years, life continues to thrive. The Loriciferans not only survive without oxygen, but they do so while being surrounded by poisonous sulfides and extreme salty water that would make normal cells turn into dried-out husks.

Such a prime example of life existing in the most challenging of Earth environments gives hope and credence to the argument that despite apparently unlivable climates on our nearest neighbors' bodies and objects spied further afield, life could still prevail against the odds. It is argued that archaea may well be the oldest form of life on Earth. Such organisms might survive on other worlds, including Venus, the past environment of Mars, Jupiter, Jupiter's moon Io, and also the planet Saturn.

Atmospheric Constituents

Earth's atmosphere consists of a mix of 78% nitrogen, 21% oxygen, 0.93% argon, 0.04% carbon dioxide, with traces of neon, helium, methane, krypton, and hydrogen, as well as water vapor.

The atmosphere is divided into five layers:

1. **Troposphere:** Distance (from ground level upwards): 0–12 km. This layer is the first above the surface of the Earth and contains half of the planet's atmosphere.
2. **Stratosphere:** Distance (above ground level): 12–50 km. A very stable layer used most frequently by air traffic to fly above weather systems, which occur at the lower troposphere level. Within the stratosphere lies the ozone layer, which absorbs harmful rays from the Sun.
3. **Mesosphere:** Distance (above ground level): 50–80 km. This layer is responsible for the burning up of meteors and other cosmic debris, as long as it is not over a certain size.
4. **Thermosphere:** Distance (above ground level): 80–700 km. Where Aurora Borealis activity can be seen, and also the layer where space vehicles such as the late lamented Space Shuttle would attain and hold orbit.
5. **Exosphere:** Distance (above ground level): 700–10,000 km. This is where the Earth's atmosphere merges into that of space. The atmosphere is extremely thin here.

The troposphere is the lowest level in the Earth's atmosphere, extending to a height of between nine km at the poles to 17 km at the equator, although there is some variation in these measurements due to weather phenomena. The troposphere is bounded above by the tropopause, a limit marked in most places by a temperature inversion (a layer of relatively warm air existing above a colder layer) and in other areas a zone that is isothermal in height (the process where there is a change of a system but the temperature remains constant).

Although variations in temperature do occur, the general rule dictates that temperature usually declines with altitude within the troposphere, as this layer is largely dependent on energy transfer from the surface of the Earth, thus decreasing in value as one gets further from the surface. Containing approximately 80% of the entire mass of the combined atmosphere levels, the troposphere is denser than all its overlying atmospheric layers, because a larger atmospheric weight sits on top if it and causes it to be severely compressed. The lower part of the troposphere alone holds 50% of the total mass of the atmosphere. The bulk of the Earth's weather also takes place in this layer, as nearly all of the atmospheric water vapor is found at this level.

The stratosphere is the second lowest layer of the Earth's atmosphere, separated from the troposphere by the tropopause. At the very height of the stratosphere, the atmospheric pressure is approximately 1/1000 of the pressure at sea level. The ozone layer is incorporated within the stratosphere, a part of the atmosphere that contains relatively high concentrations of ozone. Whereas the troposphere reflects temperatures in decreasing value upon descent, the stratosphere is a layer in which temperatures steadily rise with increasing altitude. The rise in temperature is caused by the absorption of ultraviolet (UV) radiation from the Sun by the ozone layer, which restricts turbulence and mixing, as opposed to the troposphere, which, because of the heat transfer from the ground, has vertical mixing, troposphere from the Greek word "tropos" meaning "turn." As a greater percentage of weather activity is confined to the troposphere, the stratosphere is virtually free from cloud and weather-related turbulence. This makes for stable conditions and is the desired height of airlines and other air traffic.

The mesosphere is the third highest layer of the Earth's atmosphere, where temperatures drop with increasing altitude to the mesopause that marks the top of this middle layer. Here lies the coldest place on Earth, with the average temperature down to –85 °C (–120 °F). The air is so cold that even the very scarce water vapor can be sublimated into polar-mesospheric noctilucent clouds. These clouds can be sighted with the naked eye when sunlight reflects off them. The best time to sight the clouds is either up to two hours before sunset or up to two hours before sunrise, and when the Sun is around four to 16 degrees below the horizon. This is the layer where most meteors burn up, the fragments of which are also associated with the sighting of noctilucent clouds.

The thermosphere marks the fourth highest layer in the Earth's atmosphere, extending upwards from the mesopause, which separates it from the mesosphere. Due to fluctuations in solar activity, the height of the thermosphere varies quite considerably, with its ceiling just below the most outer region of the atmosphere, the exosphere. The lower boundary of the exosphere is called the exobase or exopause, a critical altitude in the structure of the atmosphere where barometric conditions no longer apply.

The temperature of the thermosphere gradually increases with height, and whereas in the stratosphere the temperature inversion is due to the absorption of radiation by the ozone layer, in the thermosphere the tempera-ture inversion occurs due to the extremely low density of its molecules. The air is so rarefied that an individual molecule travels an average distance of one kilometer between collisions with other molecules.

The thermosphere, where the International Space Station (ISS) orbits, is completely cloudless and free of water vapor. Non-hydrometeorological phenomena such as the Aurora Borealis and Aurora Australis are occasion-ally seen in this layer. Although the atmosphere is very thin both in the thermosphere layer and exosphere layer, there is still enough air to cause a slight amount of drag force on satellites orbiting within these layers, so much so that if uncorrected, they would eventually undergo an orbital decay, which subsequently cause them to enter and burn up in the atmo-sphere as their orbit spiraled downwards. The ISS loses about two km in altitude each month, which means an upward boost from rocket engines is required to stabilize and maintain its correct orbital position.

At the very limit of our atmosphere lies the exosphere, the very outer-most layer of Earth's atmosphere. At its ceiling, the solar wind awaits. This layer is composed of extremely low densities of hydrogen, helium, and several heavier molecules, including nitrogen, oxygen, and carbon dioxide closer to the exobase. Whereas molecules traveled an average distance of one kilometer to interact in the thermosphere, here, they can travel literally hundreds of kilometers without colliding with one another. In this rather strange layer, the exosphere does not behave like a gas at all, as particles constantly escape into near space. Once free from the exosphere, the parti-cles follow ballistic-like trajectories, falling in then out of the solar wind. These trajectories take on the flight of a thrown ball or cannonball as the molecule gradually falls back to Earth under the influence of gravity. However, some particles resist the pull of gravity and continue off into space, causing a constant but miniscule atmospheric leak of molecules away from the Earth each year.

Beyond the exosphere is the vacuum of space. But in so many ways, the exosphere is nearly the same, extremely thin in nature and forming the most tenuous upper strand that merges almost unnoticed with the airless void of outer space. There is no clear boundary between them, just a gradual fading at around 190,000 km above the Earth's surface, about halfway to the Moon. At this distance, radiation pressure from sunlight exerts more force on hydrogen atoms than does the pull of the Earth's gravity. Indeed, satel-lites have detected a faint glow of UV radiation scattered by hydrogen atoms at a height of around 100,000 km, a region of UV glow known as geocorona.

The Sun in Space

At the heart of our Solar System lies the Sun. At a distance of 150 million km from Earth, this middle-aged yellow star shines its way through the depths of space to breathe life into our world. Its heat and light warm the Earth's surface, driving weather patterns, ocean currents, and the process of photosynthesis. Lying at the outer reaches of the Milky Way in one of its spiral arms, the Sun is some 30,000 ly from the center of our galaxy. At this distance, it takes about 250 million years to complete an orbit around the galaxy.

The Sun is an estimated 4.5 billion years old and has a mean radius of 696,000 km, which makes the diameter around 1.392 million km. The Sun is 109 times larger in diameter than Earth and 9.7 times the diameter of Jupiter. However, those are planetary size comparisons, which can be misleading. In fact, the Sun is actually quite small—for star size comparisons, the red giant star Betelgeuse in the constellation of Orion is 1,000 times larger than the Sun, while the largest star known, VY Canis Majoris, measures 2,000 times larger than the Sun. If VY Canis Majoris were at the center of our Solar System, it would encompass everything up to and past the planet Saturn.

The Sun's mass is 330,000 times that of Earth, 1,048 times that of Jupiter, and 3,498 that of Saturn. The most massive star (different from the largest) is Eta Carinae, which has a mass 150 times that of the Sun. The Sun's density is 1.41 (water = 1), and it has a luminosity of 3.828×10^{26} Watts.

At the Sun's core lies one of the most powerful processes in the universe, that of nuclear fusion. Hydrogen nuclei smash together, forming helium and releasing substantial amounts of energy, which in turn release light and heat from its surface. At the core, the temperature reaches 14 million °C (22.5 million °F). As long as this fuel burns, the Sun and every other star will continue to generate light and heat until the source of the fuel is eventually exhausted. The balance between the outward push of gas heated by fusion and the inward pull of gravity is called hydrostatic equilibrium.

At a point closest to the Sun's core lies the radiative zone. Here, the gas is smooth and static, the energy diffusing through the zone as radiation. Above the radiative layer lies the convective zone, where swirling currents of gas carry the Sun's energy further outward through the convection process. In this zone, gas is simultaneously heated from below by fusion and cooled from above as energy is released into space. Gas convention works in a churning motion, similar to water just before it boils.

The photosphere is the Sun's visible surface, where its atmosphere becomes transparent to visible light. Sunspots depict the cooler regions of the photosphere. The chromosphere and corona are the outermost layers of the Sun, the chromosphere being ten times hotter than the photosphere, with the corona hotter still. The corona is so hot it escapes the Sun's gravity, flowing outward into space as the solar wind. Temperature on the Sun's surface reaches 5,500 °C (9,932 °F) with sunspot temperatures measured at 4,000 °C (7,232 °F).

Our very active yellow dwarf star makes up 99.86% of the mass of the entire Solar System. Its gravity holds the Solar System together, keeping everything from the largest planets right down to the smallest debris in its orbit. Electric currents in the Sun generate a magnetic field that is carried out through space on the solar wind, a stream of electrically charged gas that extends outward from the Sun in all directions. Many phenomena are directly related to the solar wind, including geomagnetic storms than can knock out power grids on Earth, the Aurora, and plasma tails on comets, which always point away from the Sun.

The Sun Seen Through Our Atmosphere

Light rays that travel straight down do not bend, while rays that enter the Earth's atmosphere at a shallower angle get refracted, roughly following the direction of the Earth's curvature. This means that celestial objects in the zenith position overhead appear in the correct position, as the light is direct and unaltered. However, objects closer to the horizon appear to be higher up in the sky than they actually are, with one particular example a daily occurrence, that of the rising and setting of our Sun. The light rays from the Sun enter the Earth's atmosphere at a particular shallow angle, and because of refraction, the Sun may be seen for several minutes before it actually rises in the morning, and for several minutes after it has actually set in the evening.

One of the consequences of this refraction is that, contrary to popular belief, the day and night are not exactly 12 hours long on the days of Equinoxes. Because of the refraction of sunlight, daytime is slightly longer than night time on the March and September Equinoxes. This effect also has an impact on the time that the Moon can be seen before moonrise and after moonset.

A number of atmospheric phenomena occur when light is refracted by ice crystals in the atmosphere, all dependent on the size, shape, and distance from the Earth's surface. These ice crystals can be found high up in the Earth's atmosphere, scattered right across the globe and throughout the year. Generally the crystals, which have a hexagonal molecular structure, are found in cirrus clouds. Cirrus clouds, identified as thin and wispy in appearance, are the most common form of high-level cloud, typically more than 6,000 meters high. Sometimes, in extremely cold weather, the crystals form at a lower level nearer to the Earth's surface, these types of crystals being known as "Diamond Dust."

While the molecular structure of the ice crystals is always hexagonal, the actual shape and size differs, perhaps taking the form of a plate or a column. Plate-like crystals float in the air horizontally like a leaf, whereas column-shaped crystals tend to float in the air vertically. The interplay between the shape of the crystals and the angle between their facets creates the different atmospheric phenomena, and because the angle is always 60 or 90 degrees, scientists classify them as 60° or 90° phenomena.

Halos

Two kinds of halos are most commonly seen from Earth, 22° and the 46° halos. While halos can be seen throughout the year, the winter months lend themselves to a general increase in sightings, mainly because the colder weather creates much better conditions than milder weather for the formation of halo-generating ice crystals.

22° halos are formed when light passing through ice crystal bends 22°, while a similar parallel bending occurs in 46° halos.

Sundogs

Known as a sundog, a mock sun or perihelion, this phenomenon consists of glowing spots around the Sun and is classified as a member of the halo family. A sundog is created by sunlight refracting off plate-shaped ice crystals high up in the cirrus or cirrostratus clouds. Sundogs are some of the most frequently observed of all Sun-related phenomena and are observed across the world during all seasons. Two sundogs often flank the Sun with a 22° halo. Sundogs tend to be most visible when the Sun is close to the horizon.

To the observer, the sundog appears as a pair of subtly colored patches of light to the left and right of the Sun, and at the same altitude above the horizon as the Sun. Also known as phantom suns, their scientific name is parhelia, from the Greek word meaning "beside the sun." The Ancient Greeks believed that seeing sundogs in the morning meant a storm was on the way. Native Americans believed that the bright spot of light glowing beside the Sun looked like a dog leaping through a ring of fire. In medieval times, the three bright lights were sometimes interpreted as the sign of the trinity and a sign of great fortune.

Fig. 1 Sundogs. Courtesy of NASA/Nina Garcia

Sundogs in their appearance can look like bite-sized rainbows, but with a reverse color scheme. Primary rainbows are red on the outside and violet on the inside. Sundogs are red on the side nearest the Sun, with colors grading through orange to blue as they travel away from it. A double rainbow's secondary bow colors are arranged in this same way (red inside, violet outside). Sundogs are similar to secondary rainbows in another aspect: their colors are fainter than those of a primary bow. How visible or whitewashed a sundog's colors are depends on how much the ice crystals "wobble" as they float in the air. The more they wobble, the more vibrant the sundog colors will be.

The Moon also has its own moondogs, known as a mock moon or paraselene. Light from the Moon creates glowing spots on both of its sides. This phenomenon is rarer than sundogs because it only occurs when the Moon is full or very close to being full.

Sun Pillars

The appearance of light pillars is generally confined to cold climates, such as the Arctic regions. Manifesting as columns of light emanating from below or above a light source, light pillars occur when natural or artificial light reflects off flat ice crystals in the air close to the Earth's surface. Light pillars caused by the Sun are called Solar or Sun Pillars, while pillars caused by the Moon's light are called Lunar or Moon Pillars. Solar pillars are usually seen when the Sun is near the horizon.

While light pillars are normally associated with the Sun or the Moon, they can also occur due to the presence of artificial lights. Light pillars of all nature tend to take on the color of the light source.

Parhelic Circle

Parhelic circles are quite rare, appearing as white circular bands in the sky at the level as the Sun or the Moon. Parhelic circles are generated when vertical or nearly vertical ice crystals of any shape reflect sunlight or moonlight.

Generally, only parts of the circle can be seen at any time, but if complete, such a circle would stretch all around the sky. These sections of the parhelic circle are less common than sundogs and 22° halos. Chiefly white in color (because they are caused by reflection), they can at times show a bluish or greenish tone near the 120° parhelion and be reddish or deep violet along the fringes. The 120° parhelion is a rarity in itself, another optical illusion occasionally appearing along with very bright sundogs. When visible, 120° parhelia appear as white-bluish bright spots on the white parhelic circle and are the product of at least two interior reflections of the hexagonal ice crystals. These are difficult to sight, as they tend to fuse with the clouds in the sky.

Another optical phenomenon is a Liljequist parhelion, a rarely sighted halo in the form of a brightened spot on the parhelic circle, approximately 150 to 160° from the Sun, i.e. between the position of the 120° parhelion and the anthelion. The anthelion (from the Greek word meaning "opposite the Sun") is also rare in nature, appearing on the parhelic circle opposite the Sun as a faint white spot, not unlike a sundog, and may be crossed by an X-shaped pair of diffuse arcs. How anthelia are formed remains disputed.

The Liljequist phenomenon was first observed by Swedish meteorologist Gosta Hjalmar Liljequist (1914–1995) in 1951 at Maudheim, Antarctica, during the Norwegian–British–Swedish Antarctic Expedition in 1949 to 1952. While the Sun touches the horizon, a Liljequist parhelion is located approximately 160° from the Sun and is about 10° long. As the Sun rises up to 30°, the phenomenon gradually moves towards 150°, and as the Sun gradually reaches over 30°, the optical effect of the Liljequist parhelion disappears.

Liljequist's findings were subsequently simulated by Dr. Eberhard Trankle (1937–1997) and physicist Robert Greenler in 1987, and theoretically explained by Mathematics Professor Walter Tape in 1994.

Fire Rainbow

Fire Rainbows are neither fire nor rainbows but are so called because of their brilliant pastel colors and flame-like appearance. Technically known as circumhorizontal arcs, fire rainbows are a halo of ice formed by ice crystals that produce a halo so large that the arc appears parallel to the horizon. Fire Rainbows occur mostly during the summer and between particular latitudes. The conditions to form a circumhorizontal arc are very precise: the Sun has to be at an elevation of 58° or greater with cirrus clouds at high altitude, bearing the plate-shaped ice crystals. To complete the specific conditions, sunlight has to enter the ice crystals at a specific angle. This combination of ingredients makes for a rare but beautiful phenomenon.

Mirages

If the temperature of air through which light is traveling is the same, the light will pass through the air in a straight line. Mirages occur when the ground is very hot and the air is cool. The heat from the ground warms a layer of air just above the ground. When light moves through the cold air and into the layer of hot air, it is refracted. A layer of very warm air near the ground refracts the light from the sky nearly into a U-shaped bend. If a steady temperature gradient exists, light will follow a curved path towards the cooler air. Colder air has a higher index of refraction than warm air does and, as a result, photons travel through hot air faster than they can through cold air because the hot air is less dense.

One of the more unusual and more complex mirages is known as the Fata Morgana, from the Italian name for the Arthurian sorceress and powerful enchantress, Morgan le Fay. The name was so given from a belief that these mirages, often seen in the Strait of Messina, were fairy castles in the air–Fata Morgana, Italian for "Morgan the Fairy." The castles or false lands in the air were thought to be created by her witchcraft to lure sailors to their deaths. Fata Morgana is responsible for some of the more unusual and profound distortions of astronomical objects including the Sun, Moon, and planets, as well as bright stars and comets.

There is a distinct difference between mirages and other refraction phenomena. One of the most prominent features of a mirage is that it can only produce images vertically, and not sideways, whereas a simple refraction might distort and bend images in any way. There are two variants of astronomical mirages: the inferior mirage and the mock mirage.

The inferior mirage of astronomical objects is the most common mirage, occurring when the surface of the Earth or the oceans produces a layer of hot air of lower density just above the surface. This produces two images, the inverted one and the erect one, in an inferior mirage, both displaced from the geometric direction to the actual object. Strangely, while the erect image is setting, the inverted image appears to be rising from the surface. One such shape of an inferior image sighted at sunset is the "Etruscan vase," named by Jules Verne (1828–1905). Also known as an "Omega Sunset," this mirage occurs as the Sun descends towards the horizon, with a second Sun seemingly rising to greet it. Eventually, the two then join at a red-hued vertically stretched "stem." As the stem shortens and thickens, the two Suns appear like the Greek letter omega. The Suns continue to merge until eventually only a discus shape remains. On occasions, a green flash may occur.

A mock mirage of astronomical objects is more complex than an inferior image. While an inferior image of astronomical objects can produce two images, a mock mirage is capable of producing multiple miraged images, with the generated shapes constantly changing and unpredictable in nature. These are more pronounced than the inferior image. A setting Sun can be distorted and seemingly cut into horizontal slices, indicative that one or more temperature inversion layers are at work. The layers are themselves sometimes made visible by the dust and aerosol they trap, along with greater refraction. Sunlight becomes refracted by the different density layers, and sections can often be seen sinking and rising in the layers.

When there is a temperature inversion present—warm air overlaying cooler layers of air— the mirages are known as M-Mir sunsets. Sometimes the temperature and density differences are so great that rays are trapped within layers, causing ducting, where the trapped rays might travel large distances before escaping.

Mock mirages at sunset are dependent on the observer's height. The observer must be above the inversion layer but perhaps not geographically very high, because inversions can be very close to the land or sea surface. The differences in what one observer might see as opposed to another observer can, depending on height and location, be vastly different.

Mirages could explain the so-called Marfa lights, also known as the Marfa ghost lights, observed near US Route 67 on Mitchell Flat, east of Marfa, Texas. The lights have been reported as rather distant bright lights distinguishable from ranch lights and automobile headlights on Highway 67. The first published account of the sightings was in the July 1957 edition of *Coronet* magazine, although anecdotal observations date back to 1883.

Explanations of the Marfa lights include a sort of mirage caused by sharp temperature gradients between cold and warm layers of air. With Marfa located at an altitude of 1,429 meters above sea level, differentials of 22–28 °C (40–50 °F) between high and low temperatures are quite common.

Various scientific experiments have been conducted to investigate this phenomenon. Two of the investigations, one conducted in May 2004 and the other in May 2008, concluded that the lights were indeed mirages, mere atmospheric reflections, based on automobile lights and small fires. Yet, similar investigations into the Paulding Lights (also called the Lights of Paulding or the Dog Meadow Lights) appearing in a valley outside of Paulding, Michigan have been inconclusive.

Green Flash

The top of a mock mirage sunset can produce a single or even multiple flashes, generally either green or blue. The flash, which can be seen shortly before sunrise but more commonly before sunset, happens when the Sun is almost entirely below the horizon, with the barest upper edge of the Sun still visible. For just a second or two, that upper rim will appear to be the color green. It is fleeting and difficult to spot but is quite a spectacle for the lucky observer.

The green flash results from the observer looking at the Sun through an increasingly thicker atmosphere as the Sun sinks lower and lower in the sky. Water vapor in the atmosphere absorbs the yellow and orange colors in white sunlight, and air molecules scatter the violet light. That leaves the red and the blue-green light to travel directly toward the observer. Just before the sun vanishes, the red light's shallower ripple causes it to shoot overhead and miss the observer's eyes. With its steeper wavelength, green remains the sole surviving color. However, atmospheric conditions like moisture and pollution can warp and deflect the verdant tone before it reaches our eyes.

Although there appears to be no optimal condition for viewing a green flash, the chances of observing the phenomenon are heightened when viewing a horizon free of pollution. One is more likely to see the green flash when there is a visibility of several miles, almost to the point of the curvature of the Earth. Naturally, a cloudless sky is a must. Sightings of the green flash frequently occur at the ocean, where more of the atmosphere is visible and the line of sight is virtually parallel to the horizon. A common misconception for the increased numbers of sightings at sea is that the Sun shines

Fig. 2 Green Flash, observed January 4, 2011. Courtesy of Juan Jose Manzano, Grupo de Observadores Astronomicos de Tenerife

through the waves at a low angle. However, any light entering a wave is bent downwards into the water and therefore cannot escape. Prairies also have the appropriate conditions for sighting the green flash.

One of the most intriguing sightings linked to the green flash is the documented "green sun" that was observed in Antarctica. The sighting was recorded by Admiral Richard Byrd (1888–1957) and his crew on an expedition to the southernmost continent, a region that contains 90% of all of the ice on Earth in just one and half times the size of the United States. In 1931, green flashes were recorded occurring intermittently for over 30 minutes. In March 1983, sightings were also made of the phenomenon from the UK Amundsen–Scott South Pole Research Station, Antarctica, where flashes were documented lasting for an even longer length of time than those seen first by Admiral Byrd and his crew.

Admiral Byrd, who specialized in exploration and expeditions into the unknown, served as a navigator on several polar exploration flights, including the first flights over the Arctic and the North Pole, as well as Antarctica and the South Pole. In his first flight over the North Pole, Byrd claims to have discovered an entrance to Aghartha, a legendary city at the center of the Earth!

It is important to stress that with any observations made of the Sun at whatever time in the sky, care and attention is required to ensure safe viewing. It is dangerous to stare directly at the Sun even during sunrise or sunset, and this danger is only heightened when using any sort of optical aid to magnify the Sun.

The Belt of Venus

Known as the Belt of Venus, the Anti-Twilight Arch, or Venus's girdle, this atmospheric phenomenon is seen shortly before sunrise or after sunset, during civil twilight, when a pinkish glow can be seen extending roughly 10 to 20° above the horizon. Venus when visible is typically located in the Belt, hence the association. The name of the phenomenon alludes to the *cestus*, a girdle or breast-band (strophian) of the Ancient Greek goddess Aphrodite, customarily equated with the Roman goddess Venus.

The Belt of Venus is actually alpenglow, which is an optical phenomenon that appears as a reddish glow near the horizon opposite the Sun when the solar disk is just below the horizon. The effect is easily visible when mountains are illuminated but can also be seen when clouds are lit through backscatter, the reflection of particles back to the direction from which they came. Since the Sun is positioned below the horizon, there is no direct path for the sunlight to reach the mountains. Alpenglow is a result of the light being reflected off airborne precipitation, ice crystals, or particulates in the lower atmosphere.

Like alpenglow, the backscatter of reddened sunlight creates the Belt of Venus, but unlike alpenglow, the sunlight refracted by the fine particulates that cause the rosy arch of the Belt hovers much higher in the atmosphere, which makes the effect persist much longer than alpenglow. As twilight progresses, the flow is separated from the horizon by the dark band, or dark segment, of the Earth's shadow. Naturally, the shadow of the Earth is substantial, with the observer having to turn their head to see the whole aspect of the shadow. The shadow is curved in just the same way that the whole Earth is curved, extending hundreds of thousands of miles into space, so far that it can touch the Moon. Whenever this happens, there's an eclipse!

The Moon

Sighting the Moon during the daytime still surprises many who catch sight of it. As many of us have no reason to look upwards unless prompted by a noise or sudden darkening, it is of no surprise that the daytime Moon goes largely unnoticed. The Moon is not in the daytime sky all the time. Like the Sun, it spends half its time below the horizon.

A very common misconception in astronomy is that the Moon is directly opposite the Sun in the sky. In fact, the Moon is only in this position for a single instant in the whole lunar month, the exact time of a full Moon, when it is 180° away from the Sun. The rest of the month it can be anywhere from 0 to 180° away and, at least in theory, visible in the daytime sky. At full Moon, the Moon is exactly opposite the Sun, meaning that the Moon rises just as the Sun is setting. This is also the only night in the month when a lunar eclipse can occur. Even so, eclipses normally happen only one full Moon out of every six; the other times, the Earth's shadow is either too high or too low to touch the Moon.

Two major factors contribute to the Moon being visible in daylight. First, it must be bright enough that light penetrates the scattered blue light of our own sky. Secondly, the Moon must be high enough in the sky to be visible—obvious, but essential. Because of the Earth's rotation, the Moon is above the horizon roughly 12 hours out of every 24, and since those 12 hours almost never coincide with the roughly 12 hours of daylight in every 24 hours, the possible window for observing the Moon in daylight averages six hours per day. The Moon is visible in daylight nearly every day, the exceptions being close to the new Moon, when the Moon is too close to the Sun to be visible, and close to full Moon, when it is only visible at night.

Moon Illusion

Like the Sun, the Moon is more prominently noted nearer to the horizon, when its size and presence in the sky look duly enhanced. Although the angular dimensions actually remain constant, we perceive the Moon to be larger in this region. The angle that the full Moon subtends at the observer's eye can be measured directly with a theodolite to show that it in fact remains constant as the Moon rises or sets. Photographs taken of the Moon at different elevations confirm that its size remains the same.

Documented by various cultures, this illusion has spawned several theories to explain its appearance, all of which remain debatable.

The Ebbinghaus illusion is one of the more favored ways to explain the appearance of the larger Moon when nearer the horizon. It argues that if the Moon is high above the horizon, there are large expanses of space or clouds around it. When the Moon is near the horizon, there may be objects near it that appear small. These objects could be distant clouds, buildings, or trees. Distant objects in the field of vision near the Moon make us perceive the Moon as large relative to the surrounding objects.

The apparent distance hypothesis states that the perceived size of an object is related to its background. If the object is near the distant horizon, we are conditioned to think of it as larger than the same object when positioned overhead.

According to the angle of regard hypothesis, the Moon illusion is produced by changes in the position of the eyes in the head accompanying changes in the angle of elevation of the Moon.

A simple way of determining for oneself that the illusion is simply illusion is to hold a pebble that measures roughly 8.4 millimeters wide. At arm's length, approximately 640 millimeters, and with one eye closed, position the pebble so that it covers the Moon when it is high in the sky. Make sure that the pebble is not over exaggerated in size to cover the Moon—a pebble that provides an eclipse-like size will more than suffice. When the seemingly very large Moon is positioned on the horizon, you will find that the same pebble will also cover it, revealing that there has been no change in the size of the Moon.

Earthshine

When viewing the crescent Moon shortly after sunset or before sunrise, aside from the brightness of the crescent Moon itself, the observer can also clearly see the rest of the Moon as a dark disk. That pale glow on the unlit part of the crescent Moon is light reflected from the Earth, known as earthshine.

Earthshine occurs when the sunlight reflects off the Earth's surface and illuminates the unlit portion the Moon's surface. Since the light that generates earthshine is reflected twice—once off the Earth's surface and again off the Moon's surface, the light is much dimmer than the sunlit portion of the Moon. The phenomenon can also occur on the moons of other planets, then known as planetshine. During NASA and the ESA's joint Cassini–Huygens space probe mission (launched on October 15, 1997), onboard equipment used Saturn's shine to image portions of the planet's moons, even when they did not reflect direct sunlight.

Apart from planetshine, there is also ringshine, when sunlight is reflected by a planet's ring system onto the planet or onto the moons of the planet. This was observed and carefully documented in photographs sent back by the Cassini–Huygens orbiter.

Earthshine's brightness is also affected by the Moon's albedo (the measurement of exactly how much sunlight a celestial object can reflect). Measured on a scale of 0 to 1, an object with an albedo of 0 does not reflect sunlight and is perfectly dark. An object with an albedo of 1 reflects all of the sunlight that reaches it. The Moon has an average albedo of 0.12, meaning that it reflects about 12% of the sunlight that reaches it. The Earth has an average albedo of 0.30, reflecting around 30% of the sunlight that hits its surface. Because of this, the Earth when observed from the Moon would look about 100 times brighter than a full Moon that is seen from the Earth.

Earthshine is most apparent one to five days before and after a New Moon, with April and May seeing a more intense shine as a potential consequence of global warming.

Blue Moon and Red Moon

There are two definitions of a Blue Moon. The first is a seasonal Blue Moon, the third full Moon in an astronomical season with four full Moons (versus the usual three). The second is the monthly Blue Moon, the second full Moon in a month with two full Moons.

Blue Moons are rare, hence the popular phrase "once in a Blue Moon." Blue Moons happen once every two or three years. In the 1,100 years between 1550 and 2650, there will be 408 seasonal Blue Moons and 456 monthly Blue Moons. Double Blue Moons are even rarer, occurring only about three to five times in a century. The last occurrence of a double Blue Moon was in 1999; the next will be in 2037.

Other combinations of Blue Moons also exist. Between 1550 and 2650, only 20 years had or will have one seasonal and one monthly Blue Moon. The last time this occurred was 1934; the next will be in 2048. Triple Blue Moons, a combination of one seasonal and two monthly Blue Moons in the same calendar year, happen 21 times in the same time span. The last occurred in 1961, and the next will be in 2143. There can never be a double seasonal Blue Moon, as this would require 14 full Moons in the same year, which is not possible because the time between two full Moons is approximately 29.5 days.

Statistically speaking, Blue Moons make up about 3% of all full Moons. Supermoons are approximately 25% of all full Moons, and total lunar

Fig. 3 Lunar eclipse: a blood-red moon. Courtesy of NASA

eclipses occur during 5.6% of full Moons, meaning that a Blue, Super, totally eclipsed Moon occurs with 0.042% of full Moons—once every 2,380 full Moons or so. On average, that corresponds to once every 265 years!.

As far as astronomical rarities go, the rarest of all Blue Moons is the Moon that actually takes on the appearance of being blue. The Moon, full or any other phase, can appear blue when the atmosphere is filled with dust and smoke particles of a certain size, measuring slightly wider than 900 millimeters. The particles scatter the red light, making the Moon appear blue. This is known as Mie scattering and can happen after such events as a dust storm, a forest fire, or a volcanic eruption.

Several eruptions have generated Blue Moons, including Mt Krakatoa in Indonesia in 1883, Mt St Helens in the USA in 1980, El Chichon in Mexico in 1983, and Mt Pinatubo in the Philippines in 1991. All of these made the Moon look blue in appearance because of the particles driven high into the atmosphere following the outbursts. It is suggested by some that the term "once in a Blue Moon" is based on these rare occasions, rather than the definition afforded by the full Moon.

More commonly seen than a Blue Moon is a Red Moon, which generally appears due to the size of particles in the atmosphere during total lunar eclipses. On the night of the Blue Moon on January 31, 2018, there was such a lunar eclipse. So, on this particular January night, a chance arose to see a red Blue Moon. If that wasn't enough of a rarity, it was almost a Supermoon as well, earning the nickname at the time of a Super Blue-Blooded Moon. For observers in North America, it was the first time all three of these phenomena had lined up since 1866.

When a full Moon takes place when the Moon is near its closest approach to Earth—at a time when the center of the Moon is less than 360,000 km from the center of the Earth— it is called a Super Full Moon. When there is a new Moon around the closest point to Earth, it is known as a Super New Moon. A Micro Moon, on the other hand, is when a full or new Moon is near its farthest point from the Earth, around apogee, giving it the name Micro or Mini Moon, or Mini Full Moon or Mini New Moon. This takes place when the center of the Moon is farther than 405,000 km from the center of the Earth.

Moonbow

Moonbows are similar to rainbows, but they are created by moonlight instead of sunlight. This phenomenon is quite rare because of the necessary weather and astronomical conditions that must be in place for it to appear.

To witness a moonbow, the Moon has to be very low in the sky, no more than 42° from the horizon, with the phase of the Moon full or nearly full. The sky must be very dark—any bright light can obscure the phenomenon. There also have to be water droplets present in the air in the opposite direction of the Moon. As with a rainbow and the Sun, water droplets still floating in the air after rain act like prisms, scattering the moonlight into a visible spectrum of color. The colors however are absent to the naked eye, only presenting themselves on film and digital exposure. More often than not, the moonbow will appear to the observer rather grey or white in color, with stronger moonlight yielding a more brilliant display.

Along with the Hawaiian Islands, the Scottish Highlands in the United Kingdom have a good record of moonbow sightings, mainly due in Scotland's case to its wet climate. As with the appearance of rainbows, moonbows can be quite frequent in certain locations, usually where waterfalls are present to generate layers of mist in the air. The moonbows created near waterfalls are often called spray moonbows, with some of the best viewing locations including Yosemite National Park in California during the spring snowmelt.

Cumberland Falls State Resort Park in Kentucky affords a good opportunity to sight the phenomenon. At 38 meters wide and 18 meters high, the Cumberland Falls are known as the Niagara of the South, with mist at the falls creating colorful rainbows by day and milky moonbows on clear nights when a full Moon is present. The moonbows are seen when the moonlight is refracted in the mist, which is protected from wind dissipation by of the steep gorge walls.

Moonbows have been sporadically spotted over active lava flows in Hawaii and in the cloud forests of Costa Rica when Christmas winds blow in clouds of mist from late December through early February. Sightings have also been recorded by the Plitvice Lakes in Croatia.

Daytime Observation of Planets

The planets should not be discounted from observations that, while not atmospheric, make for other notable daytime phenomena.

It is possible is see a number of planets during daylight hours, most notably Venus. An ideal way to sight Venus in the daytime is when the Moon is nearby. A small crescent Moon can be an excellent way to orientate yourself in the sky, allowing you to focus on the distance of the Moon before turning further afield to Venus. Once you have the crescent Moon as a nearby guide to Venus, first observe the pairing as long as possible before sunset on the previous evening. If your sky is exceptionally clear, make a mental note of Venus' position in relation to the Moon. Make several observations of the position of the pairing so that you can sweep quite naturally between the two of them without having to scour the sky continually for Venus. If possible, use a chimney stack, prominent building feature, or any attribute that that can form a marker system so that returning to the very same spot the following day will generate the same observing conditions as the previous evening.

Without the aid of the Moon as a guide, select a time when Venus is prominent in the night sky, charting its progress as much as possible to allow you to make an accurate determination of where the planet should be during the hours of daylight. Venus makes for quite a surprisingly easy target, but knowing exactly where to look is key. Be mindful of the Sun's proximity to the planet, using extreme caution to sweep the area with binoculars or any other astronomical aid during the day.

Venus can yield some very nice phases during the day, but apart from the novelty of sighting it during daylight hours, there is not a great deal else to be gleaned that would make it preferential to nighttime observation. The same applies to the rather ghostly appearances of other planets, in particular Jupiter and Saturn. Mercury and Mars can also be found, and with the Sun as an easier target, along with the use of good planetarium software to pinpoint the planets, one could very much flip the daytime sky into a nighttime of observations.

One of the most famous sightings made of Venus during a daytime apparition was when President Abraham Lincoln spotted the planet on March 4, 1865. At the time of sighting, the streets of Washington, DC were packed with crowds watching the inauguration of the president for his second term in office. Suddenly, someone in the crowd saw something in the sky, a tiny brilliant point of light. Word of the sighting spread quickly as onlookers turned their attention to the skies. The commotion eventually reached Lincoln himself, and soon the president was also pointing at the light in the sky. Venus was in plain view, with Lincoln's personal bodyguard Sergeant Smith Stimmel recalling that the apparition was observed between midday and one o'clock.

Other naked-eye sightings of the planet during daylight hours exist in several anecdotes and records. Edmund Halley calculated its maximum brightness in 1716, when many Londoners were alarmed by the appearance of Venus in the daytime. Napoleon Bonaparte (1769–1821) once witnessed a daytime apparition of Venus while at a reception in Luxembourg.

Aside from Venus, the other planets do make for quite a challenge to spot. A more seasoned background in astronomy affords the best chances, along with good atmospheric conditions and accompanying good eyesight. The best time to spot Jupiter in the daylight is when it's near a "quadrature," about 90° away from the Sun in the sky. This is similar to the arrangement of the first quarter and last quarter Moon. At quadrature, the sky is slightly darker at 90° to the Sun, due to a phenomenon known as polarization. As with Venus and the crescent Moon, it is of great help to have a quarter Moon nearby as an astronomical signpost or guide.

As far as astronomical rarities are concerned, the sighting of planets, while not overly scientific in nature, can make for another area of astronomy in its entirety. Should the chance present itself to observe the daylight planets, the opportunity should be seized, not least because only a minority of observers has ever seen them in this out-of-context state, as opposed to the traditional backdrop of a night sky.

Other Atmospheric Phenomena

Besides the illusions of the Sun and the Moon, favorable atmospheric conditions can allow for the observing of comets and sometimes bright meteors during daylight hours, and even the occasional glimpse of a bright star like Sirius.

A rarity presents itself during a total solar eclipse, when the chance to see bright stars and planets makes for unusual viewing. During the eclipse of August 21, 2017, dubbed "The Great American Eclipse," such an opportunity arose to see both Venus and Jupiter, which were positioned nearly halfway across the open sky at the time. Mars and Mercury were positioned much closer to the Sun, but with the Sun eclipsed, there was a favorable chance to sight both. There were opportunities not only to see at least four planets but also to glimpse Sirius, Capella, Castor and Pollux, and Regulus.

Incredibly, from different observing points, artificial satellites were also seen, with one location in Oregon, USA observing one of the Flock-2 Earth-monitoring satellites, appearing as a point of light where Mercury was positioned. During the eclipse, the International Space Station (ISS) made three passes through the Moon's penumbra, experiencing a partial solar eclipse each time. The ISS didn't pass through the Moon's umbra and therefore didn't experience a total solar eclipse from orbit. During the first pass through the umbra, the ISS experienced a partial solar eclipse of 38%, with 43.9% covered on the second sweep. Pass number three saw 84% of the Sun covered.

Unexplained Atmospheric Rarities

While many atmospheric phenomena and rarities can be explained, there are some that remain unexplained or questionable, one such case being the Hessdalen lights, which have been observed in the 15-km Hessdalen valley in rural Norway, home to around 150 residents.

The Hessdalen lights are of unknown origin, appearing during day and night and apparently seeming to float through and above the valley. The lights are usually bright white, yellow, or red in appearance and can appear above or below the horizon. They can be on view for just a few seconds or for up to two hours, with the lights sometimes seen to be traveling at enormous speeds, or at times swaying slowly back and forth, or on other occasions simply hovering in midair.

The lights have been reported in the valley since the 1930s, with especially high activity recorded between 1981 and 1984. During that period, lights were being observed 15 to 20 times per week, attracting many visitors to the area to see the spectacle. As of 2010, sightings have dwindled, with only 10 to 20 made yearly.

So, what are they? Are they atmospheric in nature? Scientists believe that they could possibly be the result of a natural "battery" source buried deep underground, created by metallic minerals reacting with a sulfurous river running through it.

Dr. Jader Monari from the Institute of Radio Astronomy in Medicina, Italy made a lengthy study of the Hessdalen site in 1996. At that time, Monari discovered rocks in the valley that are rich in zinc and iron on one side of the river running through it, and rich in copper on the other side.

By taking rock samples and creating a miniature valley of his own, Monari found that electricity flowed between the two rocks, generating power that could light a lamp. Monari surmised that bubbles of ionized gas were being created when sulfurous fumes from the River Hesja reacted with the humid air in the valley. The geology of the valley also forms electromagnetic field lines, which could explain why the light orbs move around.

Ball Lightning

Usually concomitant with thunderstorms, ball lightning is literally a blazing ball of lightning, a mobile burning sphere capable of passing through blocks of metal, wood, and even buildings. Accompanied by a strong sulfurous odor, with a lifespan of one second to one minute, ball lightening remains largely unexplained. It has been documented as transparent, translucent, multicolored, evenly lit radiating flames, filaments, or sparks, with shapes that vary between spheres, ovals, tear-drops, rods, and plates. Red, orange, and yellow are the most reported colors.

This dangerous atmospheric phenomenon can be pea-sized or several meters in appearance. Given its unpredictability and infrequency, ball lightning's formation and constituents remain questionable. What descriptions of ball lightning exist vary quite widely. It has been described as hovering, moving up and down, sideways, or taking totally unpredictable trajectories, moving with or indeed against the wind.

The appearance of ball lightning is associated with power lines, altitudes of 300 meters and higher, thunderstorms, and also ironically, much calmer weather.

Amongst the many and varied explanations for ball lightning is the Black Hole Hypothesis. This hypothesis suggests that some ball lightning is the passage of microscopic primordial holes through the Earth's atmosphere! Several names are connected with the idea, including Leo Vuyk in 1992 and Leendert Vuyk 1996. The first detailed scientific analysis was published by American physicist Mario Rabinowitz in the *Astrophysics and Space Science Journal* in 1999.

Earthquake Lights

These lights are often seen before the onset of an earthquake, and sometimes even after, comprising bright flashes of white and blue light, localized to places with intense tectonic disturbances. Earthquake lights have a lifetime varying from just a few seconds to around 10 minutes. Records of the phenomenon date back to 373 BCE, with more recent video footage coming from the 2009 L'Aquila earthquake in Italy and similar accounts in 2010 and 2011.

Many explanations have been proposed to explain this unusual luminous aerial phenomenon, but no definitive answer exists. One such proposal suggests that the generation of earthquake lights involves the ionization of oxygen to oxygen anions by breaking of peroxy bonds in some types of rocks due to the high stress before and during an earthquake. After the ionization, the ions travel up through the cracks in the rocks, whereupon at higher altitudes they can ionize pockets of air, forming plasma that emits light.

Chapter 9

The Noisy Universe

There is much to be heard in our Universe if you only know where to listen.

The idea of audible astronomy perhaps springs to mind the subtle background static, that fuzzy haze of nothingness, which forms the cosmos' very own elevator music.

The sounds generated from the universe are as much a part of astronomy as observational work. From the echoes of the beginnings of time to the dullest thud of a meteorite striking the Earth's ground, every sound generated by our cosmos teaches us an important lesson about our existence and the world around us.

Human-Made Sounds

Many of the sounds we hear in the sky are artificial in nature. Sputnik 1's famous chiming in space was sent back to Earth, awing listeners from every nation. Never before had such a dull and ordinary sound signaled something so momentous. For the world, the sound meant far more than simply the fact that Sputnik 1 was properly transmitting. It chimed the arrival of our first concrete foothold in space—no matter whose foot it was—and exemplified the daring feats that had been and could be accomplished by humankind.

© Springer Nature Switzerland AG 2018

J. Powell, *Rare Astronomical Sights and Sounds*, The Patrick Moore
Practical Astronomy Series, https://doi.org/10.1007/978-3-319-97701-0_9

Sputnik, meaning "Fellow traveler," launched on October 4, 1957, heralding the start of the Space Age and the dawn of such signals over the decades that followed. Sputnik 1's chief constructor was Mikhail Stepanovich Khomyakov. The final design measured 525 mm in diameter, comprising two 2-mm-thick hemispheres hermetically sealed with O-rings, connected by 36 bolts, encased inside a 1-mm-thick heat shield. Weighing 83.6 kilograms, Sputnik carried two pairs of antennae made up of two whip-like parts, 2.4 meters and 2.9 meters in length. The antennae, designed by a team led by Mikhail V. Krayushkin, were formed to give an equal distribution of signal with equal power in all directions, so that despite the rotation of the satellite in flight, the reception of the transmitted signal would be independent of Sputnik 1's motion.

Sergei Korolev was the leader of the team that developed the booster rocket on which the satellite gained space entry. Korolev based a fair proportion of the design on the theoretical work of Russian school teacher Konstantin Tsiolkovsky, who in 1903 had written about the possibilities of multistage boosters fueled by liquid hydrogen and oxygen. His application of the work culminated with the successful launch of Sputnik 1 from the Baikonur Cosmodrome in early October 1957, sending off the satellite—which would eventually burn up in the atmosphere on January 4 1958—to spend three months in orbit around the Earth.

During its time, Sputnik 1 completed 1,440 orbits of the Earth, traveling 70 million km at a speed of 29,000 km per hour, taking 96.2 minutes to complete each orbit.

Despite this monumental achievement, it was the satellite's time in orbit and Sputnik 1's audible beeps that truly captured the world's imagination. Powered by a one-watt, 3.5-kg radio unit developed by Vyacheslav I. Lappo from the Moscow Electronics Research Institute, the radio unit sent signals from the craft that were transmitted back to Earth. Sputnik generated two signals: one on 20.0005 MHz, which transmitted in 0.3 second pulses, and the other, operating on 40.002 MHz, emitting pauses of the same duration filled by pulses.

Sputnik's steady beep reached listeners around the globe, sparking amazement wherever it was heard. News reports encouraged anyone with a short-wave radio receiver to tune into the craft as it passed over the area within which anyone possessing a radio lived. The American Radio Relay League (ARRL), issued its own instructions to hear the *beep, beep* sound, with the first recording of Sputnik 1's signal done by RCA engineers near Riverhead, Long Island. The recording was subsequently played on NBC radio, with students at Columbia University's WKCR FM station also making their own recording and broadcasting the sounds generated by the satellite. Despite the initial agreement between the Soviet Union and the USA to use equipment that was universally compatible so that all could hear Sputnik 1, the Russians changed their mind, introducing lower frequencies. Sputnik's signals continued for 21 days until the transmitter batteries finally died on October 26, 1957.

The power supply for the entire Sputnik 1 output came in the form of three silver-zinc batteries, developed at the All-Union Research Institute of Current Sources (VNIIT) under the leadership of Nikolai S. Lidorenko. Two of the three batteries were installed and dedicated to powering the radio transmitter; the third controlled the temperature regulation system onboard the satellite, which kept a close eye on Sputnik 1's heat output. A series of switches and a fan were put in place to assist with any anomalies and make necessary adjustments to influence temperature values.

With Sputnik 1 whizzing around in orbit, the Earth not only listened but watched for the craft via a network of observers organized through the citizen science project entitled Operation Moonwatch. With the enlisted help of 150 observing stations in the United States and other countries around the globe, alerts went up as Sputnik 1 was sighted in the dawn and evening skies. As it made its historic passage overhead, all binoculars and telescopes turned its way.

The success of Sputnik 1 was soon followed by Russian cosmonaut Yuri Gagarin (1934–1968), who on April 12, 1961 made the first human-crewed orbital flight. US astronaut Alan Shepard (1923–1998) made the first human-piloted spaceflight on May 5, 1961. Colonel John Herschel Glenn Jr. (1921–2016) was the first American to orbit the Earth, circling it three times in 1962. On March 18, 1965, Soviet cosmonaut Alexei Leonov made the first spacewalk in history, with American astronaut Edward Higgins White II (1930–1967) emulating the task on June 3, 1965.

Beyond humankind's own first transmitted noise from space from the antennae of Sputnik, there have been other sounds from space that have both thrilled and terrified their captive audience. For a wider audience via the medium of both radio and television, Neil Armstrong (1930–2012) stepped down onto the lunar surface at 02:56 UTC on July 21, 1969 and said the unforgettable words, "That's one small step for man, one giant leap for mankind." Despite the USA's ongoing competition with the USSR, the occasion marked a great triumph of the species in which hopefully all nations could find a common ground.

One must not forget the cost to human life in such far-reaching endeavors. The Space Shuttle Challenger disaster on January 28, 1986, humbled and reminded the world of the great odds they faced in the pursuit of such endeavors.

Another such lesson came with the plight of the crew of Apollo 13, where audio of the crew's hardships and ultimate triumph was transmitted and heard across the world over the nerve-wracking course of this mission gone awry.

Apollo 13 was the seventh manned mission in the successful Apollo program conducted by NASA. The goal for its three crew members— Commander James Arthur Lovell Jr., Command Module Pilot John Leonard (Jack) Swigert (1931–1982), and Lunar Module Pilot Fred Wallace Haise— was to touch down on the lunar surface and explore the Fra Mauro formation. They were never to make the lunar surface.

Launched from Kennedy Space Center, Florida, on April 11, 1970, at 19:13 UTC, Apollo 13 comprised two independent spacecraft that were joined by an interlocking tunnel. The two craft were the orbiter Odyssey and the lander Aquarius.

Disaster struck just two days into the mission following the now-infamous oxygen tank explosion onboard Odyssey, with a call to abort the mission raised on April 14, 1970, at 03:07 UTC, 55 h 54 min 53 sec after launch. At that time, Apollo 13 was 321,860 km from home. Following the explosion, Odyssey was temporarily abandoned, the crew residing in Aquarius for much of the journey. But Aquarius wasn't supposed to be used until the landing phase of the mission, and unlike Odyssey, it did not have a heat shield, thus, the crew inevitably needed to return to Odyssey in order to make it back to Earth.

Despite limited power and lack of heat and water, the crew worked around the clock. A similarly industrious effort was being made on the ground by Mission Control, which during this time maintained the audio umbilical cord that stretched from the Earth to the stricken craft. The first human-made sounds from space on this occasion heralded news of the explosion, followed by two more significant audio markers in this eventful six-day mission. The second crucial sound came from the craft after it had rounded the far side of the Moon when, upon marking the farthest that humans had ever traveled in space, the static was broken as Apollo 13 re-emerged from the deathly blanket of static.

Closest approach to the Moon occurred on April 15, 1970, at 00:21 UTC at a distance of 253.7 km. With the Earth a colorful, distant disc in a sea of black, the journey home could begin.

The third key sound broke yet another barrier of static, as communications were lost during Apollo 13's descent to Earth through the atmosphere. The tired crew—in the case of Haise, suffering from a kidney infection— had powered up Odyssey and prepared for splashdown in the Pacific Ocean. Odyssey was in poor condition, and with doubts over the integrity of the heat shield, there was much that could go wrong. The static and ominous silence was broken on April 17, 1970, at 18:07 UCT, as splashdown was achieved, signaling the end of a traumatic but triumphant mission. All three crew members were safely picked up 45 minutes later by the USS Iwo Jima.

When the life of satellites expires, many are left to drift in orbit, adding to the vastly growing quantity of manmade space junk. The junk is the inevitable graveyard to the continued exploration of space, an unfortunate but necessary byproduct of pushing back the boundaries.

On September 5, 1977, a Titan IIIE/Centaur launch vehicle blasted off from Cape Canaveral in Florida. Onboard was Voyager-1, which was to make a journey that earned it the accolade of farthest human-made object from Earth, nearly 18,000,000,000 km away. At that distance, light takes over 16 hours to reach Earth from Voyager-1.

Voyager-1 has also been traveling away from both the Earth and the Sun at a relative speed faster than any human-made object at a velocity of 61,400 km/h, hurtling on a trajectory that will see the probe arrive at Proxima Centaur in 73,600 years. If its current speed is maintained, Voyager-1 will take about 114,000 years to cover a light year.

Voyager-1 and its sister craft Voyager-2 carry our very own "message in a bottle," a Golden Record Disc attached to the outer body of the craft. On the disc are 115 images and our very own entry into the Noisy Universe, a collection of natural sounds found on Earth, from the noises made by surf to that of wind and thunder. Other sounds for any finder of the disc include those of birds, whales, and other animals. There is also a selection of music that lasts 90 minutes—Eastern and Western classics and a variety of ethnic selections, and of course a greeting from humankind, purveyed in spoken word from Earth-people in 55 languages.

The greeting messages recorded on the disc start with Akkadian. Akkadian is the extinct language of the Akkad, written in cuneiform (wedge-shaped marks on clay tablets). It has two dialects, Assyrian and Babylonian, widely used from about 3500 BCE. It is the oldest Semitic language for which records exists. The greetings end with Wu, a modern Chinese dialect of which had over 80 million native speakers in 2007.

The Golden Record Disc is protected in an aluminum sleeve, complete with a cartridge and a needle. It has instructions in symbolic language, explaining the origin of the spacecraft with information on how to play the record.

Most of the instruments onboard Voyager-1 have been disabled. The craft entered interstellar space on August 25, 2012, becoming the first human-made object ever to do so. Along with its sister craft Voyager-2, whose outward route is different to Voyager-1, the pair conducted an epic "Grand

Tour" of the Solar System's giant planets, flying by Jupiter, Saturn, Uranus, and Neptune.

In August 2017, the 40th year anniversary since the launch of Voyager-1 and Voyager-2, both craft were still receiving command signals from Earth, but only from Australia, at the Canberra Deep Space Communication Complex. Located at Tidbinbilla and opened in 1965, the complex as of late 2016 boosts five large antennae, making it the only Earth-bound communications link left for the Voyager craft. Both craft were still responding to Canberra, generating very weak signals—only one-tenth of a billion-trillionth of a watt—but signals nonetheless!

As of September 1, 2017, Voyager-1 sat at a distance of 20.87 billion km from Earth, more than 139 times the distance from our planet to the Sun. On the same day, Voyager-2 was 17.17 billion km distant from home.

By 2030, both craft will be out of power, and there the story would seem to end as our own contribution to the Noisy Universe at last goes silent.

Noises from the Natural World

Discovery in the field of radio astronomy is not just confined to professionals. Over the decades and with increasing levels of input into the field, the role of the amateur, as with observational astronomy, has become significantly enhanced. The scientific community has quite rightly recognized that the level of sophistication of amateur instruments now allows both professionals and amateurs to work alongside each other in a two-layered approach.

You can never have too many people watching and listening to the skies, as what one might miss may subsequently be heard by another or by a few whose collective findings once correlated produce a dynamic and insightful picture. Further, the "burden" held by professional astronomers to undertake more precise, in-depth, and ultimately time-consuming work can be lessened by amateur input, which allows professionals with more intricate observational and listening tools to prioritize more specialized tasks.

Both radio astronomy and optical astronomy examine the electromagnetic radiation originating from outside of the Earth's atmosphere. The basic differences between the two fields are the method and tools used to detect this radiation and the length or frequency of the waves they study. Light and radio waves are both manifestations of the same energetic phenomena. However, because radio waves are much longer than optical waves, the telescopes used to detect them must be significantly larger than optical telescopes.

Radio astronomy was born in the early 1930s, when American physicist and radio engineer Karl Guthe Jansky (1905–1950) first discovered radio waves emanating from the Milky Way. Jansky, working for the Bell Laboratories, was trying to determine the origin of a noise that was showing up in receivers operating in the 20 MHz region of the radio spectrum.

Jansky built a steerable antenna and began searching for the source of the noise by taking directional measurements. To Jansky's surprise, he discovered that the noise was from an extraterrestrial source! Jansky published his discoveries, expecting positive feedback that could lead to a much more in-depth undertaking to pinpoint the source. However, the overall astronomical community dismissed the findings as either irrelevant or just curious at best. Fortunately, not all of the community was flippant about Jansky's discoveries, with electronics engineer Grote Reber (1911–2002) carefully reading through the published work. Reber speculated that the signals Jansky was hearing were in fact of thermal origin, caused by very hot objects, and as such should be easier to detect at higher frequencies. Since the original work was conducted at 20 MHz (about 15-meter wavelength) and a beam width of about 25 degrees, Reber, a keen amateur radio enthusiast, wanted to narrow the effective beam width in order to refine the detail.

In order to achieve this, Reber reasoned that he would need to build a receiver and antenna considerably more advanced than Jansky's. In the summer of 1937, Reber used his own resources, funding, and boundless enthusiasm to build the first parabolic reflector radio telescope.

The telescope was massive, measuring 9.5 meters in diameter, focusing to a radio receiver eight meters above the dish. This was a fine achievement, made even more so by the fact that Reber built the telescope without any financial help or construction assistance. The entire assembly was built on a tilting stand, allowing it to be pointed in various directions, though not turned. Work on the telescope was finally completed in September 1937. The term "Radio Telescope" had not yet been coined, but we can certainly attribute the first building of such a scope to Reber. For nearly a decade after the telescope's construction in 1937, Reber was the world's only radio astronomer.

Reber was unable to prove his hypothesis with his first receiver, which operated at 3300 MHz, as the receiver failed to detect signals from outer space. Undeterred, he built another receiver that operated on 900 MHz; this too failed to bring any results. For Reber, it was a case of third time lucky, with a 160-MHz receiver proving successful in 1938, confirming what Jansky had originally discovered.

Reber went on to detail the first radio map of the galactic plane, completing a radio frequency sky map in 1941 and making further observations that extended the map in 1943. His published work included the concept of "Cosmic Static," and his data was published as contour maps showing the brightness of the sky in radio wavelengths, also revealing the existence of radio sources in Cygnus A and Cassiopeia A for the first time.

It was this search for static and noise that led to the development of radio telescope astronomy, which expanded to become just as important as the observational branch.

Astronomical phenomena have certain sound characteristics, a mixture of signal properties such as frequency, phase, amplitude, and in some cases, specific repetitive patterns. It is possible from these signals to produce a mathematically assembled "radio picture" of these cosmic objects, allowing us to build a pictorial representation of the object from its sound that can be used to complement studies done through the telescope.

As one can imagine, a great deal more work goes into the production of an image from a radio telescope than one that can be taken instantly through optical means. For a single dish telescope, such as the Parkes 18-meter Dish stationed in New South Wales, Australia, the telescope scans across an object and receives radio waves from each little point across the object being studied. Some of the points in the object emit more energy than others. Each pixel is individually stored, with the computer converting this information into a series of numbers. If radio waves are weak at any particular position on the object, a small number would be recorded in the pixel, with no radio wave received registering a zero in the pixel. A higher concentration of radio waves would thus reflect a higher number in the pixel, which astronomers then apply different colors for to distinguish different areas of intensity. Gradually, they build up an overall picture of the object. The process can take hours, perhaps days, with the final data process possibly running into weeks, depending on the object and how many pixels and associated colors it commands.

Extraterrestrial radio signals are extremely weak, so weak that if all the signal energy ever received (from sources other than the Sun) from all the radio telescopes ever built around the world were combined, the collective energy output would not even be enough to melt a solitary snowflake.

The radio telescope captures a wide area of signals and focuses them into a much smaller field, much in the same way that a reflecting optical telescope operates. The term "radio optics" refers to this similarity. Since the word "light" means electromagnetic radiation in scientific jargon, all the same basic equations and formula, theories and principles can be applied to radio, infrared, and visible light. Optical telescopes operate at much higher frequencies and microscopic wavelengths, while radio telescopes work at significantly lower frequencies and longer wavelengths. Resolution in optical astronomy can be paralleled by beam width in radio astronomy.

Comprised of a sensitive radio receiver and an energy measuring device, the large antennae seen on most telescopes are built large in order to make their "beam patterns" as small as possible. The beam pattern is the two-dimensional area projected upon the celestial sphere, to which the telescope will be sensitive. A small beam pattern allows the telescope to resolve the level of signals arriving from regions separated only by a small angular distance. Multiple antennae are sometimes combined as "arrays" to enhance resolution. Widely separated antennae may have their signals combined in an "interferometer" arrangement, where resolutions can be obtained that surpass those of optical telescopes.

An array of separate telescopes, mirror segments, or radio telescope antennae works together as a single entity to provide higher resolution images of astronomical objects. While found in optical astronomy, the widest use of the astronomical interferometer is in radio astronomy, in which signals from radio telescopes are combined. A mathematical signal processing technique known as aperture synthesis combines the separate signals to create high resolution images. At shorter wavelengths, it is more difficult to combine light from separate telescopes, because the light must be kept coherent within a fraction of wavelengths over optical paths, requiring very precise optics.

At optical frequencies (blue-green light 600,000 GHz or a wavelength of .0005 mm) a 1-meter perfect mirror will have a beam width of approximately .00003 degrees. The same mirror operating at radio frequencies (30 GHz for example, with a wavelength of 1 cm) will have a beam of approximately six degrees.

Pulsars and Quasars

Some cosmic objects produce signals that can be considered as a point source, such as pulsars and quasars. There are many other radio sources in the universe, all of which carry a signature unique to its sender. The task of the telescope is to receive all the sound and noise from the object and its vicinity, using fine tuning and intricate analysis to weed out unrelated sounds in order to get a true sense of what is being heard. Only from this filtered sound alone can any determination be made involving the radio properties and sole identity of the object.

Pulsars and quasars were at the forefront of popular radio astronomy when this particular branch of the science began to break new boundaries, and accessibility to the field became wider than was previously possible—though one should be reminded that the first radio telescope was built by Grote Reber in his backyard!

Pulsars are a type of neutron star or white dwarf, the dead relic of a massive star. What sets pulsars apart from regular neutron stars is that they're highly magnetized and rotate at enormous speeds, emitting a beam of electromagnetic radiation. They were discovered in 1967 by astrophysicist Dame Susan Jocelyn Bell Burnell, at the time a postgraduate student. The discovery was heralded as one of the most significant scientific achievements of the 20th century. Her thesis supervisor Antony Hewish and astronomer Martin Ryle jointly won the Nobel Prize in Physics. Sadly, Bell (later to become Bell Burnell) was excluded, despite being the first to observe and precisely analyze the pulsar.

These neutron stars, which consist mainly of neutrons, are very small, around 20 to 24 km in diameter, but have enormous mass, between 1.1 and three times the mass of our Sun, which is approximately 1.392 million km in diameter. Having such a large mass squeezed into a small volume means that their density is incredibly high. A cubic centimeter of neutron star material would weigh about 500 million tons. The gravitational pull on the surface of such an object would be about one billion times stronger than the gravitational pull on the surface of the Earth.

When observed from Earth, pulsars appear as stars blinking at intervals, creating a constant rhythmic pattern. Pulsars radiate two steady, narrow beams of light in opposite directions. Although the actual light from the beam is steady, pulsars appear to flicker because they also spin. As the pulsar rotates, the beam of light may sweep across the Earth, then swing out of view, then swing back around again. To an observer on the ground, the light comes and goes in and out of view, creating the impression that the pulsar is blinking on and off. Whereas a lighthouse generates a steady light beam from a constant defined angle, a pulsar's beam is not typically aligned with its axis of rotation.

Because the "blinking" of a pulsar is caused by its spin, the rate of the pulses also reveals the rate at which the pulsar is spinning. This rate of spin can categorize such stars as "slow pulsars," or faster spinners as "millisecond pulsars," rotating at hundreds of times per second. Some 200 millisecond pulsars have been discovered.

Telescope Contributions and Discoveries

Fermi Gamma-Ray Telescope

NASA's Fermi Gamma-ray Telescope, launched on June 11, 2008, is a powerful space telescope that has allowed astronomers to conduct detailed observations of supermassive black-hole systems, pulsars, the origin of cosmic rays, and searches for signals of new physics. Fermi, which originally bore the named Gamma-ray Large Area Space Telescope (GLAST), had its mission renamed in honor of Professor Enrico Fermi (1901–1954), a pioneer in high-energy physics. The telescope has detected 2,050 gamma-ray-emitting pulsars, including 93 gamma-ray millisecond pulsars.

An electronvolt is the unit of energy close to that of visible light. The Fermi Gamma-Ray Telescope observes light in the photon energy range of 8,000 electronvolts (8 keV) to greater than 300 billion electronvolts (300 GeV). This allows the telescope to observe photons with energy levels thousands to hundreds of billions of times greater than what the naked eye can see.

The Fermi telescope, armed with such instruments as the Large Area Telescope (LAT) and the Gamma-ray Burst monitor, has made some staggering discoveries. Aside from playing a role in the discovery of more than 100 pulsars, it aided in the detection of giant bubbles stretching more than 25,000 ly above and below the plane of the Milky Way. The mysterious structures are thought to possibly be the result of outbursts from supermassive black holes at the center of the galaxy, or evidence of a burst of star formation a few million years ago. Together, these bubbles cover more than half of the visible sky and emit gamma rays. Radiating about the same amount of energy as 100,000 supernovae, their combined width is 50,000 ly across.

Further observations conducted from the German-led Roentgen satellite (ROSAT), launched June 1, 1990, provided hints of bubble edges close to the galactic center, with NASA's Wilkinson Microwave Anisotropy Probe (WMAP), launched June 30, 2001, detecting an excess of radio signals at the position of the gamma-ray bubbles.

The Fermi Gamma-ray Space Telescope, along with the Netherlands-based Low Frequency Array (LOFAR) radio telescope, completed in 2012 by ASTRON, the Netherlands Institute for Radio Astronomy, identified the pulsar spinning at 707 rotations per second, or 42,000 revolutions per minute, making it the second fastest known. LOFAR detected pulses from PSR J0952-0607ad (J0952 for short) at radio frequencies around 135 MHz, which is about 45% lower than the lowest frequencies of conventional radio searches.

The fastest known pulsar is PSR J1748-2446ad (J1748 for short), which was discovered by Professor Jason W.T. Hessels of McGill University on November 10, 2004. Resident in the constellation of Sagittarius, this pulsar spins at 716 times per second, or 716 Hz. The theoretical maximum for the rotational speed of a pulsar is 72,000 rpm before the pulsar literally disintegrates as it tears itself apart. J1748 spins at a rate of 43,000 rpm, 60% short of this theoretical barrier. It has been speculated that beyond the barrier, formation is impossible, and perhaps higher spinners exist but we are just unable to detect them.

J0952 is located in the constellation of Sextans at a calculated distance of between 3,200 and 5,700 ly away. J0952 contains about 1.4 times the Sun's mass and is orbited every 6.4 hours by a companion star that has been whittled away to less than 20 times the mass of the planet Jupiter. The pulsar falls into the classification of pulsars that stream matter from their companion star onto themselves and, in doing so, gradually push up the spin rate and greatly increase their emissions. If J0952 continues to follow what appears to be a natural evolutionary pattern, drawing material away, it will eventually exhaust and evaporate its companion. This type of effect has earned J0952 and similar companion systems the name "Black Widows," or "Redback Pulsars," referring to the spiders who consume their mates.

The Fermi Gamma-ray Space Telescope has made a telling impact on research into pulsars, with other major contributions to pulsar research conducted by the Arecibo telescope in Puerto Rico, the Green Bank Telescope in West Virginia, the Molonglo telescope in Australia, and Jodrell Bank in Cheshire, England.

Ground-Based Radio Telescopes

Built in the early 1960s, the telescope at Arecibo was for over 50 years the world's largest radio telescope of its kind. Located 16 km south of the town of Arecibo in Puerto Rico, the site employs a 305-meter spherical reflector that consists of perforated aluminum panels. These panels focus the incoming radio waves on moveable antenna structures positioned about 168 meters above the surface of the reflector. The antenna structures can be moved, making it possible to track a celestial object in different parts of the sky. The observatory also has an auxiliary 30-meter telescope that serves as a radio interferometer and high-power transmitting facility used to study the Earth's atmosphere.

The telescope, which sits inside a naturally formed karst sinkhole, discovered the first extrasolar planets around pulsar B1257+12 in 1992. The observatory is also responsible for producing detailed radar maps of the surfaces of Mercury and Venus, discovering that Mercury rotated every 59 days instead of the originally calculated 88 days, so it did not always show the same face to the Sun.

The first binary pulsar was also discovered at the observatory by American astronomers Russell Hulse and Joseph Hooton Taylor Jr., earning them the Noble Prize for Physics in 1993.

In 1974, the Arecibo telescope attempted to communicate with potential extraterrestrial life. Known as the Arecibo Message, the transmission was aimed directly towards the globular cluster M13, around 25,000 ly distant. The 1,679-bit pattern of 1s and 0s defined a 23 x 723-pixel bitmap imager that included numbers, stick figures, and chemical formulas, along with a very crude image of a telescope. The message, created by Dr. Frank Drake with help from Carl Sagan (1934–1996) among others, was transmitted at a frequency of 2,380 MHz and modulated by shifting the frequency by 10 Hz. The broadcast lasted less than three minutes.

Drake was instrumental in the founding of SETI (Search for Extraterrestrial Intelligence). He developed the famous Drake equation, a problematic argument used to estimate the number of active, communicative extraterrestrial civilizations in the Milky Way galaxy.

FAST Radio Telescope

The Chinese built the Five-hundred-meter Aperture Spherical Telescope (FAST) located in Pingtang County, Guizhou Province in southwest China and launched on September 25, 2016, following a construction time frame that spanned from 2011 to early July 2016. It is a novel design, using an active surface made of metal panels that can be tilted by a computer to help change the focus to different areas of the sky.

Housed in a natural depression, FAST is the world's biggest radio telescope, with a dish measuring a staggering 502 meters in width. Consisting of 4,450 triangular panels, FAST is by far the largest single-aperture telescope in the world (although arrays that link up multiple radio dishes cover more ground). The project cost $180 million. It is hoped that FAST will shed light on the universe's early days, detecting low-frequency gravitational waves and hunting for signals that may have been produced by ancient civilizations. The telescope joined the Breakthrough Listen SETI project in October 2016, adding significant assistance in the search for extraterrestrial communications. Its addition to the network increases the likelihood of discovering an alien civilization by five to ten times what it was prior to FAST becoming a reality.

Among FAST's innovations is the adoption of an active rather than passive primary reflector. The dish's reflectors can be adjusted to account for signal deformation. The result equates to a sensitively twice that of the Arecibo Telescope, the observable area of which is around 20 degrees from the zenith. FAST, with its adjustable reflector, achieves 40 degrees.

Designed, developed, and built by Chinese scientists, FAST will also survey neutral hydrogen in distant galaxies, detect faint pulsars, and probe interstellar molecules.

Green Bank Telescope

The Green Bank Telescope (GBT), or Robert C. Byrd Green Bank Telescope, is the world's largest fully steerable radio telescope. The scope honors the name of Senator Robert C. Byrd, who pushed funding of the telescope through Congress. It was built between 1991 and 2002, with first light on August 22, 2000. Considered to be a very accurate and precise telescope, its suite covers 100 MHz to 100 GHz in frequencies, with processors capable of spotting nanosecond timing differences in data.

The telescope's 2.3-acre dish surface is excellent for picking up weak radio signals from space. However, it is its steerability that sets the GBT apart from any other telescope, allowing it to cover 80 percent of the sky. Stationary single-dish radio telescopes must make do with the swath of the universe that passes directly overhead. For example, the Arecibo telescope can observe about 33 percent of the sky. The GBT's high coverage makes it a world leader in this area, and a much-needed monitor of the skies via radio waves.

Molonglo Observatory Synthesis Telescope

The Molonglo Observatory Synthesis Telescope (MOST) is located near Canberra in New South Wales, Australia, with its construction beginning in 1960 under Emeritus Professor Bernard Yarnton Mills (1920–2011). Following an upgrade, it became known as UTMOST, capable of studying the sky in millisecond timescales.

UTMOST consists of a 778-meter-long parabolic cylindrical antenna array. The array contains a total of 352 independent antennae, with 7,744 ring antennae at the focus of the parabola. The main thrust of work undertaken by UTMOST involves its examination of Fast Radio Bursts (FRBs)—short, bright, highly dispersed pulses of radio waves that occur randomly. The bursts typically last just a few milliseconds but register about a billion times brighter than anything ever observed in the galaxy. Bursts of this nature were first detected at the Parkes radio telescope by Duncan Lorimer and student David Narkevic in 2007 as they trawled through data collected from the telescope six years prior. There in the archives, was a burst that occurred on July 24, 2001, lasting less than five milliseconds and emanating from the Small Magellanic Cloud. Thus was the FRB, the Lorimer Burst FRB 010724, discovered.

Parkes Radio Telescope

The Parkes radio telescope is located 20 km north of the town of Parkes in New South Wales, Australia. Famous its work in radio astronomy among other things, the scope was one of several radio antennas used to receive live, televised images of the Apollo 11 Moon landing on July 20, 1969. Completed in 1961, it was the brainchild of Edward George (Taffy) Bowen (1911–1991), a Welsh physicist who made a significant contribution to the development of radar and played key role in the establishment of radio astronomy in both Australia and the United States. Via the persuasion of two influential contacts, Dr Vannevar Bush, and Dr Alfred Loomis, Bowen managed to gain the backing and funding of the Carnegie Corporation and the Rockefeller Foundation. In 1962, armed with a $250,000 grant, Bowen began the construction of the Parkes radio telescope.

A smaller 18-meter dish antenna was transferred from Fleurs Observatory (Mills Cross Telescope location, 40 km west of Sydney) to Parkes. Much work has been conducted on FRBs and related signals known as perytons. These perytons, named after a mythical creature by novelist Jorge Luis Borges, were found to result from the premature opening of a microwave oven door at the observatory! The microwave was discovered to release a frequency-swept radio pulse that mimics an FRB as the magnetron (high-powered vacuum tube that generates microwaves) turns off.

The 64-meter dish is not actually fixed to the top of the tower—its own 1,000 tons simply weigh it down and hold its position. Because of its large surface, the dish has to be "parked" or "stowed" in winds that exceed 35 km/h. This is executed by turning the dish to point directly upwards.

It takes 15 minutes for the dish to complete a 360° rotation, and five minutes to get to its maximum tilt of 60°. Parkes radio telescope, along with the Green Bank telescope, was used by researchers at SETI when they launched Project Phoenix, the world's most sensitive and comprehensive search for extraterrestrial life. Rather than concentrating on a large section of the sky, Phoenix narrowed its listening to around 1,000 Earth-like stars within a distance of 200 light years.

Jodrell Bank

Established in 1945 by English physicist and astronomer Sir Bernard Lovell (1913–2012), Jodrell Bank has played an important role in the research of meteors, quasars, pulsars, masers, and gravitational lenses, along with the tracking of space probes at the start of the Space Age.

The main telescope at the observatory is the 76.2-meter Lovell Telescope, the construction of which was completed in 1957. At the time, the telescope was the largest steerable radio dish telescope in the world, but after the building of the Green Bank telescope in the United States and the Effelsberg telescope in Germany, it is now the third-largest. Originally called the "250-foot telescope" or the "Radio Telescope at Jodrell Bank," it was renamed the "Mark I" in 1961, with a view to creating Marks II, III and IV. Finally in 1987, the Mark I was rather fittingly renamed after Sir Bernard Lovell, who had built the earlier Transit Telescope at Jodrell Bank in the late 1940s.

The Transit Telescope was a 66-meter parabolic reflector zenith telescope that could only point upwards. The telescope consisted of a wire mesh suspended from a ring of 7.3-meter scaffold poles, which focused the radio signals to a focal point 38 meters above the ground. Despite the telescope only being able to look upwards, it was possible to tilt the mast by small amounts in order to change the direction of the beam. The Lovell Telescope replaced the Transit Telescope, with the Mark II built on the same location.

The Mark II is an elliptical radio telescope with a major axis of 38.1 meters and a minor axis of 25.4 meters. Aside from its use as a standalone telescope, the Mark II has also been used as an interferometer with the Lovell Telescope. Constructed in 1964, the Mark II was followed by the Mark III, built in 1966. The Mark III was the same size as the Mark II and was constructed to be transportable, however, it was never moved, being decommissioned in 1996.

Whereas construction was achieved for the Mark II and Mark III telescopes, plans to build the Mark IV, V, and VA were shelved, with progress only being made up to the level of building concept models.

Both the Lovell Telescope and the Mark II contribute to the Multi-Element Radio Linked Interferometer Network (MERLIN), which comprises an array of radio telescopes spread across England and the Welsh borders. MERLIN's central point of organization is at Jodrell Bank. Five other radio telescopes at Cambridge, Defford, Knockin, Darnhall, and Pickmere complete the array. Operating at frequencies between 151 MHz and 24 GHz, MERLIN's longest possible baseline of effective aperture is 217 km. At a wavelength of six cm, MERLIN has a maximum resolution of 40 milli-arcseconds, approximately 20 times better than can commonly be achieved by the best ground-based telescopes, and comparable to the Hubble Space Telescope.

Very Large Array

The Karl G. Jansky Very Large Array (VLA) is situated in central New Mexico on the Plains of San Agustin between the towns of Magdalena and Datil. The VLA is composed of 28 25-meter radio telescopes (including one that is a spare) deployed in a Y-shaped array. All the equipment, instrumentation, and computing power function as an interferometer. Each telescope has eight receivers tucked inside. The first antenna was put in place in September 1975, and the complex was officially inaugurated in 1980. It was the largest configuration of radio telescopes in the world. Four times a year, a specially designed rail truck called a transporter picks up telescopes and hauls them one at time farther down the purpose-built track.

Situated 80 km west of Socorro, the VLA stands at an elevation of 2,124 meters above sea level. The multi-purpose instrument is designed to allow investigations of many astronomical objects, including radio galaxies, quasars, pulsars, supernova remnants, gamma-ray bursts, and radio-emitting stars, along with black holes and our own Sun and planets in the Solar System.

Amongst its impressive list of findings, the VLA has discovered ice on Mercury; a new type of astronomical object called the "microquasar;" pebble-sized chunks that show the first stages of planet formation; and, most intriguing of all, a billion-light-year-diameter "hole in the universe."

Amateur Observations

For the amateur astronomer, a whole new world is opened up through radio astronomy, beyond straightforward optical observations. Amateur radio telescopes are much harder to find and usually require a great deal of time to assemble, plus the ability to troubleshoot potential software issues. This should by no means deter anyone interested in radio astronomy.

All radio telescopes consist of three basic components:

1. Antenna
2. Receiver
3. Output Recorder

At entry level in the United States, there are three types of radio telescopes to consider:

1. **The Itty-Bitty Telescope (IBT):** A simple form of radio telescope that can be used for satellite tracking.

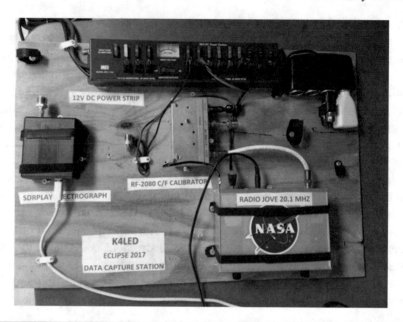

Fig. 1 Jove Radio. The radio equipment used by Larry Dodd during his observation of a total solar eclipse. Courtesy of NASA

2. **Radio Jove:** The Radio Jove Project is a NASA program aimed at Solar and Planetary Radio Astronomy for schools and the general public. The project, which began in 1998, monitors storms on Jupiter, solar activity, and galactic background noise. It incorporates advice on the building of radio telescopes (a fair amount of backyard space is required for the setup of this particular scope) and the remote use of other radio telescopes through the internet.
3. **SuperSID:** This type of radio telescope allows the amateur to collect real solar activity data from the ionosphere.

In the United Kingdom, several radio astronomy groups exist, including the British Astronomical Association–Radio Astronomy Group and the UK Radio Astronomy Association (UKRAA).

Although the amateur radio astronomer can go it alone with the research, assembly, and construction of a telescope, as with optical astronomy, seeking out a likeminded group of enthusiasts is highly recommended, given that 40% of all those involved with radio astronomy at this level are also radio ham operators.

Frequencies

Frequencies below 15 MHz or so are rarely used, due to absorption of these waves by the ionosphere. At the upper end of the frequency range, limitations are imposed by the technology needed to receive signals with such tiny wavelengths. Almost all amateur radio telescopes fall between 18 MHz and 10,000 MHz.

The exact choice of frequency for a given amateur will depend on the technical abilities of the experimenter, the type of observations being sought, the radio interference pattern in the area, and the amount of room available for the antenna.

Frequencies as designated by the International Telecommunications Union (ITU):

3 Hz-30 Hz
Wavelength: 10^8m-10^7m
Band: Extremely low frequency (ELF)

30 Hz-300 Hz
Wavelength: 10^7m-10^6m
Band: Super low frequency (SLF)

300 Hz-3 KHz
Wavelength: 10^6m-10^5m
Band: Ultra low frequency (ULF)

300 Hz-3 KHz
Wavelength: 10^6m-10^5m
Band: Ultra low frequency (ULF)

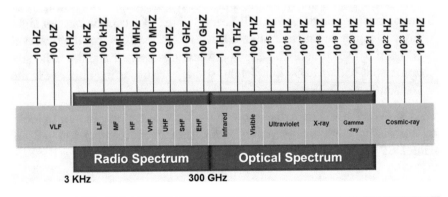

Fig. 2 Spectrum band designators and bandwidths

Objectives

Beyond the prospect of finding signals from alien life, amateur radio astronomy can provide an extensive spectrum of targets to listen to:

1. **The study of Jupiter's noise storms.** Jupiter is a source of powerful natural radio waves that can produce exotic sounds when picked up on Earth using simple antennas and shortwave radio. The sounds are naturally produced by plasma instabilities in Jupiter's magnetosphere. These plasma waves give rise to the laser radio signals that travel away from Jupiter's magnetic poles in cone-shaped beams. The beams rotate with Jupiter every nine hours and 55 minutes, making Jupiter something of a slow-turning pulsar. When the beams sweep past Earth, listeners can pick up the Jovian radio bursts in the shortwave bands between 15 and 40 MHz.

2. **Detecting ionospheric effects.** With solar flares come blasts of X-rays. When X-rays hit the Earth's ionosphere (the charged particle layers in our atmosphere), the way the ionosphere reflects radio waves is disturbed. At short-wave frequencies, a dip in signal strength of distant stations can often be observed. At VLF (Very Low Frequencies) below 150 KHz, the opposite effect is observed, with signal strength of distant stations suddenly jumping, followed by a slow decline. Using a VLF receiver permanently tuned to a distant station is a reliable way to detect X-ray flares. Most radio receivers will not tune low enough in frequency to be used for VLF solar flare detection. One can purchase an up-convertor to listen to these lower frequencies using a standard short-wave receiver. A good and inexpensive candidate for a VHF solar flare receiver is an aircraft band radio, which covers the 120–140 MHz range.

3. **Meteor detection.** Most meteors are not seen, but heard. Find a frequency where no nearby FM station is broadcasting, the best chance of which will be scanning the low-frequency end of the FM band, below 91.1 MHz. If you are watching a meteor display while monitoring your radio, most of the time you will hear a "ping" of reception but not sight any corresponding meteor streak. Most of the meteors that the observer is "hearing" are roughly halfway between the observer and the radio station, about 650 to 1,050 km distant, so they are occurring either near the horizon or just below it.

As knowledge and expertise grow in the field, attention can turn to looking for HEPs (high-energy pulses from the galactic center) and pulsars using DSP (digital signal processing).

Interpreting Sound

Space is a vacuum, meaning that it cannot carry sound waves like air does here on Earth. Still, space missions and their associated probes have heard some quite bizarre noises. NASA's Juno spacecraft captured the apparent "roar" of Jupiter, with the bow shock of crossing Jupiter's immense magnetic field recorded by the craft during a two-hour period on June 24, 2016.

Plasma waves, like the roaring surf of the ocean, were recorded by NASA's EMFISIS instrument onboard the Van Allen Probes. The probes also picked up strange whistles resulting from plasma waves interacting with the Earth's own magnetic fields.

NASA's Cassini spacecraft picked up eerie sounds from Saturn, generated by auroras near the poles of the planet. These auroras are similar to Earth's Northern and Southern Hemisphere lights. Voyager also noted evidence of lightning deep within Saturn, with radio waves yielding a static-like crackle captured by Cassini in 2006. In 2016, static was recorded by Cassini's Radio and Plasma Wave Science instrument as it crossed the plane of Saturn's rings on December 2016.

In order to better understand radio signals, astronomers sometimes convert the signals into sounds, a technique called "data sonification." On June 27, 1996, the Galileo spacecraft made the first flyby of Jupiter's largest moon, Ganymede, with an audio track record from Galileo's Plasma Wave Experiment instrument. Similar sounds were recorded when Galileo passed by Jupiter's moon Europa, resulting from the interaction of Europa and Jupiter's magnetospheres. Callisto also produced comparable sounds but appeared to show the weakest interaction with Jupiter's magnetosphere of any of the four largest moons.

During a February 14, 2011 flyby of comet Tempel 1, an instrument on the protective shield on NASA's Stardust spacecraft was pelted by dust particles and small rocks, the recording of which was relayed back to Earth.

Perhaps it is the sound generated as Voyager 1 crossed into interstellar space that conjures up the most awe-inspiring thoughts. Whilst the sound recorded wasn't real-time—indeed, it was an audio version of a graph of Voyager's Plasma Wave Science Observations of several months spanning 2012 to 2013—it nevertheless marked a turning point. For astronomers, it marked the Voyager's exit from our Solar System's heliopause, the area in which the pressures from the outside of our Solar System force what's left of the Sun's solar wind to turn back.

Chapter 10

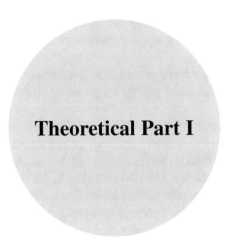

Theoretical Part I

Continuing to Question

Many theories abound regarding what could and what should exist in our cosmos. Speculation turns to fact only once enough concrete evidence has been made to mark a real discovery. Plenty of the modern concepts that scientists have put forth contain holes or unanswered questions that force us to continually question and revisit our models of the universe.

Supernovae: Creators of Black Holes

Black holes are created when the center of a very massive star collapses in upon itself. The process of collapse occurs when a large star literally runs out of fuel, causing it to break down as the evolutionary cycle of the body comes to a final and quite dramatic end. This final stage, which results in a nova, from the Latin word for "new," produces a massive outward explosion and creates a temporary new bright star that can be of great brilliance for a time, shining with the potential brightness of 10 billion suns before fading away.

A supernova is generated at the end of the life of much larger stars, the brilliance much more noticeable even in daylight. First coined by German astronomer Walter Baade (1893–1960) and Swiss astronomer Fritz Zwicky

© Springer Nature Switzerland AG 2018   153
J. Powell, *Rare Astronomical Sights and Sounds*, The Patrick Moore
Practical Astronomy Series, https://doi.org/10.1007/978-3-319-97701-0_10

(1898–1974), a supernova expels much of what was left of the dying star in a gigantic outburst, sending debris outward at velocities of up to 30,000 km/sec, or 10% the speed of light. As the material is propelled, a shockwave resonates outward, sweeping up an expanding shell of gas and dust to form a supernova remnant.

Only three such naked-eye events have been observed in our own Milky Way during the last thousand years, though many others have been observed using telescopes.

SN 1006 was a supernova that is likely to have been the brightest observed stellar event in recorded history, reaching an estimated –7.5 visual magnitude, exceeding by 16 times the brightness of Venus. Appearing in the constellation of Lupus between April 30 and May 1, 1006 CE, the sighting was recorded by observers across China, Japan, Iraq, Egypt, and Europe. There are also some North American petroglyphs possibly showing the event. Some reports state that the supernova was clearly visible during daylight hours. Astronomers calculated the distance of the supernovae to have been 7,200 ly.

As mentioned in Chapter 2, On July 4, 1054, SN 1054 was first observed, remaining visible for around two years. The event was recorded by Chinese astronomers. References to the event were recorded in a 13th-century Japanese document from the Arab world. There are also some European accounts of the supernova, but some doubt over their validity remains. SN 1054 became the famous Crab Nebula in Taurus. At its heart sits a pulsar, formed as a result of the exploding star.

In 1604, SN 1604, also known as Kepler's Supernova, occurred in the constellation of Ophiuchus at a distance of 20,000 ly. The resulting new star, Kepler's Star, was brighter at its peak than any other star in the night sky, with an apparent magnitude of –2.5. Records show that the star was visible during the daytime for three weeks, with sightings documented from Europeans, Chinese, Korean, and Arabic sources. Johannes Kepler (1571–1630) had been observing the night sky during this time and noted that he thought he had discovered a new star. Indeed, Kepler had, as it was the new star generated from the supernova. It is extraordinary to think that light years away, at any given time, a star perhaps dozens of times the mass of our Sun may simply go nova in a violent end to its life, an event that can be over in a fraction of second.

In order to understand a supernova, one has to look at the makeup and composition of the star from which it is eventually born. Stars convert hydrogen into fusion at their core, generating light and heat. The reaction releases energy in the form of photons, and this light pressure pushes against the force of gravity trying to pull the star in upon itself. Our Sun doesn't have the mass to support fusion reactions with elements beyond hydrogen and helium. So, once all the helium is exhausted, the fusion reactions cease, and the Sun becomes a white dwarf and starts to cool down. Our Sun however is relatively young, part of a generation of stars known as Population I, which are relatively rich in elements heavier than helium. An older generation of stars is called Population II, and an earlier generation of Population III stars may have existed. Our Sun has used up about half of the hydrogen fuel it needs to exist, but still has sufficient gas in the tank to last another five billion years. The Sun is classed as a main sequence star, together with 90% of all other stars in the universe.

If you have a star with eight times or more the mass of our Sun, it can fuse heavier elements at its core. When this larger star exhausts its hydrogen supply, it switches to helium in order to continue its existence. When all the helium is used up, it switches to carbon, with this process slowly making its way up the periodic table of elements until the star eventually gets to iron.

By this time, the fusion reaction takes more energy than it produces. When this occurs, the outer layers of the star begin to collapse inward in just a fraction of second, detonating as a Type II supernova. If the original star had more than 25 times the mass of the Sun, the same process occurs, but the force of the material falling inwards collapses the core into a black hole.

Stars even more massive, in excess of 100 times the mass of the Sun, explode without trace. It is believed that shortly after the Big Bang, there were possibly stars that were far greater than 100 times the mass of the Sun, perhaps greater than 1,000 times, all made from pure hydrogen and helium. Their life would have been short, erupting with apocalyptic explosions.

Aside from Type II supernovae, there are Type I supernovae, which are somewhat rarer and are created by different means. A Type Ia supernova describes the strange relationship between binary stars. One star in the pair is a white dwarf, the other the long-dead remnant of a main sequence star. The companion does not necessarily have to be a distinct star type—it could be as large as a red giant, or even another white dwarf. The crucial factor lies in the fact that the two are close enough for the one star to continual steal matter from its partner. The eventual theft of matter reaches a critical point once at an accumulated transference of 1.4 times the mass of the Sun, resulting in the theft star going supernova and completely vaporizing itself. This critical level is known as the Chandrasekhar Limit.

This limit was first indicated in papers published by German-Estonian astrophysicist Willhelm Robert Carl Anderson (1880–1940) and British theoretical physicist Edmund Clifton Stoner (1899–1968) in 1929. It was named the Chandrasekhar Limit after Subrahmanyan Chandrasekhar (1910–1995), an Indian astrophysicist who independently discovered and improved upon the accuracy of the calculation in 1930.

The 1.4 ratio has greater and more far-reaching consequences in astronomical terms, as it is used as "standard candles" to measure distances in the universe. Since astronomers know how much energy any nova has detonated with, they can calculate the distance of the explosion. A standard candle is a class of astrophysical objects such as supernovae that have known luminosity due to some characteristic quality possessed by the entire class of objects. This method contributes to the cosmic distance ladder, also known as the extragalactic distance scale, whereby alongside the standard candle, other formulae are used to calculate the distance of celestial objects.

From nova to supernova to hypernova. A hypernova, or "collapsar," is the result of what is termed an extreme core-collapse scenario. In the case of a hypernova, a star with a mass exceeding 30 times that of the Sun collapses to form a rotating black hole emitting twin energetic jets and surrounded by a disc of accretion. The explosion from a hypernova releases more energy in seconds than our Sun will in its entire ten-billion-year lifetime. Whilst extremely rare, they do occur, with the rate of a hypernova occurring in the entire Milky Way estimated to be one every million years. 25 million ly from Earth in another galaxy, astronomers have found what appears to be the remnants of a giant hypernova.

Black Holes and White Holes

From the ultimate demise and death of a star, the new life of a black hole can emerge—a strange and intriguing phenomenon around which a great deal of mystery still circulates.

In April 2018, news broke of research that suggested that as many as a dozen black holes may lie at the center of the Milky Way, supporting the widely held theory that "supermassive" black holes at the centers of galaxies are surrounded by many smaller ones.

A Columbia University-led team of astrophysicists made the discovery of the black holes, which are gathered around Sagittarius A* (Sgr A*), a supermassive hole right at the center of our galaxy. As a part of extensive research, it was discovered that Sgr A* was surrounded by a halo of gas and dust that provides a perfect breeding ground for the birth of massive stars, which live, die, and could eventually turn into black holes. In addition to the black holes that may well form naturally following the death of a star, it is possible that outside of the halo, smaller black holes fall under the influence of supermassive black holes (SMBHs). As the smaller holes lose energy, they get drawn in, eventually being held captive by the SMBHs.

While a suspected majority of black holes that fall under the influence of a SMBH remains isolated, some capture and bind to a passing star, forming a stellar binary. In a search to find X-rays emitted from these binary systems, the team, led by Astrophysicist Chuck Hailey, sifted through archival data collected from the Chandra X-ray Observatory (a Flagship-class space observatory launched by NASA on July 23, 1999), hunting down possible X-ray signatures of black hole–low-mass binaries in their inactive state. They discovered 12 within three ly of Sgr A*. Black holes in an isolated state emit little or no detectable X-rays, but in a mated state like a stellar binary, X-ray bursts are detectable—weak, but consistent and steady.

Further investigation analyzed the properties and spatial distribution of the identified binary systems and concluded that there could well be anywhere from 3,000 to 5,000 black hole–low-mass binaries and about 10,000 isolated black holes in the area surrounding Sgr A*.

Composition of a Black Hole

A black hole is an incredibly dense region of space from which nothing can escape, not even light itself. A beam of light has to cover a greater distance near a black hole. The boundary region between the black hole and the space outside it is known as the event horizon. The boundary is the point where the escape velocity from the gravitational field of the black hole is equal to light.

Due to this massive gravitational field, clocks run slower the nearer you advance towards a black hole, although the actual slowing effect is regulated and dependent on the curvature of space around the black hole. However, to an observer in that extreme gravitational field, light must appear to always be 300,000 km/s—time has to slow for the direct observer, as compared to someone else outside the gravitational pull.

The idea of a black hole was first conceived in 1783 by amateur astronomer John Michell (1724–1793), long before Albert Einstein's theory of relativity. Michell, an English clergyman and natural philosopher, published a paper for the *Philosophical Transactions of the Royal Society of London*, which was read on November 27, 1783. The paper proposed that there were such things as black holes, which Michell referred to as "dark stars." Michell suggested that there might be many of these "dark stars" in the universe, and that were astronomers to make an attempt to look for them, they should start by looking for star systems that behaved gravitationally like two stars, but where only one star could be seen. Sadly, many of Michell's contemporaries did not buy into his theories on black holes, with the memory and great insight of the man only resurrected when his works resurfaced in the 1970s. Around the same time, French mathematician Pierre-Simon Laplace (1749–1827) suggested the same idea in his 1796 book *Exposition du Systeme de Monde*. Laplace propounded the concept of black holes, suggesting that there could be massive stars whose gravity is so great that not even light could escape from their surface.

Stephen Hawking's (1942–2018) greatest and most memorable work was in response to the rather simple question, "Do black holes emit heat?" Hawking had determined that black holes adhere to the second law of thermodynamics, which meant that entropy (a measure of disorder) only increases over time. Based on the fact that anything that has entropy also has a temperature, Hawking used a significant amount of mathematics along with combined insights from Einstein's theory of relativity and quantum mechanics to take the temperature of a black hole. His theory states that black holes should have the ability to thermally create and emit subatomic particles until they are completely depleted of energy.

Hawking explained how the strong gravitational field around a black hole can affect the production of matching pairs of particles and antiparticles, as is happening all the time in empty space, according to the quantum theory. If the particles are created just outside the event horizon of a black hole, then it is possible that the positive member of the pair may escape, observed as thermal radiation emitting from a black hole, while the negative particle may fall back into the black hole. Gradually, the black hole would lose its mass. Thus, Hawking was able to show that black holes can—like just about every object in our Universe—shrink and die.

In order to better understand the possible workings of a white hole, we must first examine the three types of black hole. Black holes are categorized according to their mass. These are Stellar, Supermassive, and Miniature black holes.

1. **Stellar:** The largest, formed when an extremely giant star collapses.
2. **Supermassive:** These exist at the center of most galaxies, including our own Milky Way. They can have the equivalent mass to millions of Suns. While it is not exactly known how supermassive holes form, it is likely that these are the byproduct of galaxy formation. Because of their location at the center of galaxies close to many tightly packed stars and gas clouds, supermassive black holes continue to grow via a steady intake of matter.
3. **Miniature:** Holes with a smaller mass than our Sun. Miniature black holes have not yet been discovered. It has been suggested that these holes may have formed shortly after the Big Bang. Very early in the life of the universe, the rapid expansion of some matter might have compressed slower-moving matter enough to contract into black holes.

A white hole is exactly the opposite of a black hole. Whereas you can never escape a black hole, you can never enter a white hole. The white hole is a theoretical time reversal of a black hole, and whilst a black hole acts as a vacuum, drawing in any matter that crosses the event horizon, a white hole acts a source that ejects matter from the event horizon. While such a concept is considered bunkum by some quarters of the scientific community, its properties and capabilities remain open to speculation.

It has been speculated that in some such cases, a wormhole or Einstein-Rosen bridge may exist; rather than connecting two points in space, it would connect two points in time. Wormholes were first theorized in 1916, though the term wasn't coined until later. While reviewing German physicist Karl Schwarzschild's (1816–1916) solution to Albert Einstein's (1879–1955) field equations, which describes a particular form of black hole known as the Schwarzschild black hole, Austrian physicist Ludwig Flamm (1885–1964) theorized that another solution was possible and went on to describe what would later be called a white hole. Also credited with the concept of white holes is Russian (and former Soviet) theoretical astrophysicist and cosmologist Igor Dmitriyevich Novikov. Novikov purported the theory of their existence in relation to the so-called Schwarzschild vacuum.

Further speculation followed, stating that perhaps a white hole lies at on the other side of a black hole, where all the matter the black hole sucks up is blown out in some alternative universe. Some even argue that what we think of as the Big Bang was in fact the result of such a phenomenon.

Flamm (the 20th century Austrian physicist) noticed that the two solutions describing these two different regions of space—the one on the side of the black hole, and the other on the side of the white hole—could be mathematically connected by a kind of space-time conduit, with the black hole being the entrance and the white hole being the exit.

Could the existence of white holes therefore help us realize the potential for time travel? Could a body enter a black hole and travel through a wormhole to be subsequently ejected on the other side of the white hole, ending up in a completely different region of space at a completely different time?

There are of course many challenges to the wormhole idea. Some argue that in reality, hole itself is incredibly small, limiting exactly what it could allow through. Should a person enter a black hole, the forces applied on the traveler could be likened to that of the rack, a torture device used whereby the unfortunate soul is literally pulled apart. If a person were to cross the event horizon, their body would be literally shredded. To escape its pull, they would need to travel faster than light, which is impossible. The pulling force towards the center, known as the tidal force, would act on their head, presuming that the person went into the black hole headfirst. The force on their head would be much stronger than the gravity acting upon their toes, making their head accelerate far faster than their toes, eventually stretching the body until it was torn apart. All that actually exists in the center of a black hole is an extremely large amount of matter that has been crushed into an infinitely small amount of space. The wormhole would likely take on a very different route, one that would be highly unstable; indeed, according to Stephen Hawking inserting even a particle into the wormhole might destabilize it, leading to its collapse and rendering it useless. This theory is known as the Chronology Protection Conjecture. In essence, this hypothesizes that the laws of physics are such as to prevent time travel on all but sub-microscopic scales.

The flagstone of the white hole idea rests with Einstein's general theory of relativity, and in particular the explanation of how gravity in the universe operates. Take for example the fact that scientists have already sent muon particles (similar to electrons) forward in time, achieved by manipulating the gravity around them. This negatively charged subatomic particle, about 200 times heavier than an electron, was accelerated to close to the speed of light using the CERN muon storage ring. The accelerated muon's lifetime increased almost 30 times that of a muon at rest. The time dilation effects suggest that the muon was able to travel into the future, since it continues to exist at a future time. The result of sending a muon around the CERN storage ring loop not only confirms time dilation, but also the twin paradox that clocks sent away and coming back to their initial position are slowed with respect to a resting clock.

Along with understanding the different types of black hole, one must also have an understanding of thermodynamics. Thermodynamics is based essentially on a set of four laws, the fourth of which is untested. However, while these can all apply black holes, the second law cannot be applied to a white hole.

1. The internal energy of an isolated system is constant.
2. Heat flows spontaneously from a hot body to a cool one. Heat cannot spontaneously flow from a colder location to a hotter location. One cannot convert heat completely into useful work. Every isolated system becomes disordered in time.
3. As a system approaches absolute zero, all processes cease and the entropy of the system approaches a minimum value. (Entropy is the thermodynamic quantity representing the unavailability of a system's thermal energy for conversion into mechanical work).
4. Conservation of time. All of space, matter, and energy is contained by time like a massive bubble. Time is infinite and extends in all directions through a range of 1,440 decimal degrees simultaneously and is flexible yet destructible. It is a paradoxical element of nature but integral to its existence. Without time, there can be no gravity, no matter, no space, and consequently no energy.

The second law of thermodynamics states that the amount of entropy in the universe can only stay the same or increase. We know black holes are great at increasing entropy. However, a white hole does the exact opposite: it decreases the entropy by running the process backwards, directly going against the second law.

White holes also fall foul of the laws of physics. The whole concept of the white hole simply violates the accepted model for the entire workings of the universe. If one did exist, it would be for an incredibly short time only, quickly collapsing to form a black hole. This idea has some credence. In June 14, 2006, an incredible gamma-ray burst (GRB) was detected by NASA's Neil Gehrels Swift Observatory. Named after its principal investigator, the satellite detected a GRB burst that was subsequently labeled 06064. Lasting 102 seconds, the burst was recognized as emanating from a distant galaxy, some 1.6 billion ly distant. Usually, such bursts only last for a few seconds. So what could it have been? The burst did not fit into the normal parameters of these phenomena, since such processes take place in regions of low star formation or are commonly linked with supernovae. Later referred to by some as a "hybrid burst," GRB 06064 was immensely powerful, trillions of times more powerful than our Sun. But the question remained: what caused it to happen? Since the burst did not originate from a supernova, debate was rife, with a white hole touted as a possible explanation. What astronomers know of white holes would certainly fit the criteria—a sudden and massive outburst of matter in a magnificent burst of energy, then silence as what could have been an infinitesimally minuscule speck erupted and then vanished!

Spool backwards 13.7 billion years when all the matter in the universe was compressed down into a single, incredibly small point. This single point then enlarged in a heated explosion, an explosion that continues to expand today, with galaxies constantly moving away from us. Cosmologists have suggested that a white hole could well have been the catalyst or instigator behind the initiation of the Big Bang, as it would explain how such a gigantic amount of energy and matter just spontaneously appeared, much like the GRB 06064 outburst but with an undetermined length of existence. In this theory, all the original matter collected by a black hole would have been consumed, funneled, and then ultimately ejected by the white hole as the Big Bang. But this intriguing line of enquiry does not stop there. One question leads to further ones about the origin of the matter consumed by the black hole. Was this scenario singular, or part of a series of events carried out by many black holes?

In centuries past, much has been written about the existence of certain bodies and astronomical phenomena based on mathematics alone. Take as an early example of mathematical predictions the work of French mathematician Urbain Jean Joseph Le Verrier (1811–1877) and the corresponding discovery of the planet Neptune.

The planet Uranus was discovered by astronomer William Herschel (1738–1822) in 1781. This led to astronomers noting deviations in the orbit of the newly found planet, based on predictions made by Newton's theory of gravitation. Initially, nobody really considered the possibility that perhaps another body existed beyond Uranus, which, via its gravity, causing these deviations. Speculation and interest grew over the years following the discovery as the deviations in the orbit of Uranus became more pronounced. However, it was not until the early 1840s that Uranus' odd orbital movements really came to the fore. Suggestions of an eighth planet became commonplace—either that, or Newtonian gravity would have to be modified at large separations in order to account for the deviations.

Strangely enough, this rather bizarre suggestion of a modification to the application of Newtonian gravity could be considered the starting point for the debate around the existence of "dark matter," for here we have a perfect instance of matter not detected through its light, but through the influence of its gravitational effects!

On June 1, 1846, Le Verrier, who was an expert on celestial mechanics, published a calculation that explained the discrepancies in the orbit of Uranus by predicting the existence of a trans-Uranian planet. In the publication, Le Verrier gave precise calculations as to where he though this other planet should be located in the sky. Armed with Le Verrier's calculations, German astronomer Johann Gottfried Galle (1812–1910) and his student Heinrich Louis d'Arrest (1822–1875) discovered Neptune from the Berlin Observatory on the night of September 23, 1846. Much to the joy of Galle and d'Arrest, the new planet had been proven, with Le Verrier's calculation just one degree away from Neptune's actual location. The discovery came on the same night that Galle received communication from Le Verrier detailing the planet's calculated location.

However, Galle (who also discovered Saturn's C-ring in 1838) and d'Arrest were about to have their famous discovery challenged. News traveled quickly, reaching English astronomer George Biddell Airy (1801–1892), Astronomer Royal at the time. He swiftly came to the stark realization that in the fall of 1845, he had seen a similar calculation to that of Le Verrier, produced by English mathematician John Crouch Adams (1819–1892). For several years, Adams had been working on calculations for the location of a potential new planet, communicating his work to Airy as well as physicist and astronomer James Challis (1803–1882), director of the Cambridge observatory. Sadly for Adams, Airy and Challis thought the calculations were unconvincing, and a search of the sky based on Adams' predictions was not instigated.

The discovery by Galle and d'Arrest based on the mathematical calculations of Le Verrier had a far different reception. Challis hurried to locate the new planet based on Crouch's calculations. He failed to find Neptune, but that did not prevent a heated exchange between the English and French astronomical communities, each claiming that via mathematical means, they had the right to claim Neptune as their own discovery. Later, Airy produced documents written by Crouch that deviated one degree from that of Le Verrier, perhaps sealing his prediction in history, with credit afforded to both Le Verrier and Crouch.

Although the mathematical calculations in this example cannot be paralleled to the potential discovery of a white hole, surely the same principles for a discovery—be they intentional or unintentional—still apply?

The Moon—Earth's Offspring?

Following the creation of the Solar System and the birth of the planets, our very own Moon came into being at a much later date, some hundred million years after the Earth was born. Is it possible that the Moon was once a part of the Earth, its very own offspring? Three main theories exist to explain how the Moon came into being.

Giant Impact Theory

The theory that has attracted the most support is that the Moon was formed when the Earth was impacted upon during its early evolutionary stage, resulting in a sizeable chunk being gouged out, which later formed the Moon. We know that the early stages of planetary creation saw great turbulence, with numerous collisions taking place as the leftover clouds and dust began to bind under gravitational forces.

Many bodies attracted enough material to form planets, eight of which now make up our Solar System. Others formed, then subsequently broke up during violent collisions. The biggest and most evident result of a planet that could well have formed, but for some reason failed to, is the concentration of rubble and remains circling in the Asteroid Belt. Here, perhaps with influence from the planets that were establishing themselves on either side of the belt, gravitational pull prevented material from binding to form a new planet, instead ripping that very body apart.

Scientists believe one particular body was responsible for causing the forced separation of matter from the Earth in order to form the Moon. Known as Theia—a hypothesized protoplanet (a large body of matter orbiting the Sun that has not yet formed into a planet)—this Mars-sized body is thought to have collided with the Earth, producing energy 100 million times larger than the event that saw the extinction of the dinosaurs, if the extinction did come from an extraterrestrial body, and not that of volcanism.

Theia, derived from the name of the mythical Greek Titan and mother of Selene, the goddess of the Moon, is thought to have collided with the Earth in what some quarters of the scientific community call the "Big Splash," approximately 4.5 billion years ago. Also known as the "Theia Impact," this was probably the last great impactor event to hit the Earth. Late Heavy Bombardment (LHB) was a time when smaller impacts occurred from lesser bodies much later on, some 4.1 to 3.8 billion years ago, at a time corresponding with the Neohadean and Eoarchean eras of Earth.

The impact threw vaporized chunks of the young planet's crust into space. Gravity then bound the material together to form our Moon, the largest in relation to the host planet that exists in our Solar System, with Theia at the core of the material. This formation method would go a long way to explaining why the Moon is made up predominantly of lighter elements and is less dense than Earth, as the material that formed it came from the crust, leaving the Earth's rocky core untouched.

Samples brought back from the Apollo missions to the Moon suggest that the Earth and the Moon are very similar. This seems to go against the popular impact theory, which predicted that Earth rock and lunar samples would have vastly different compositions.

Tangential to the Giant Impact Theory is the hypothesis that the Earth's Moon formed inside a cloud of molten rock, perhaps before the Earth itself formed. The cloud that would have supported the Moon's formation is known as a "synestia," a huge, hot, doughnut-shaped mass of molten rock that forms in the aftermath of a protoplanet collision, the same sort of collision that the supposed Theia and the Earth were involved in.

It is possible too that for a brief time, the Earth became a synestia around 4.5 billion years ago. Synestias, if they actually exist, are short-lived objects, with the Earth likely to have stayed in the synestia phase for approximately a century or in the immediate period after its collision with Theia. This synestia state was broken when the Earth lost sufficient heat to condense back into a solid object.

Co-formation Theory

This theory states that moons can form at the same time as their parent planet. In such a model, gravity would have caused material in the early Solar System to draw together to form the Moon at the same time that it bound particles together to form the Earth. This co-formation would explain why the Apollo Moon rock samples showed similar compositions. It would also explain the Moon's present location. However, it would not explain why the Moon is less dense than the Earth, for if both were formed around the same time and exposed to the same intake of material, both would have started with heavy elements are their core.

Capture Theory

Is it possible that the Moon was a rogue body, moving through the Solar System and simply captured by the gravitational field of the Earth? We know that other moons have been caught in similar fashion, such as the Martian moons Phobos and Deimos, rogue bodies captured by the planet from the Asteroid Belt. The capture theory would explain the different geological makeup of the Earth and the Moon. However, the paths of other moons in our Solar System don't tend to line up with the ecliptic of their parent planet, while our Moon does. This theory would also not account for the Moon's spherical shape, which the captured Phobos and Deimos certainly don't conform to.

While it is likely that Jupiter has caught a fair share of bodies as well due to its massive gravitational field, these are for the most part small in nature and situated in the outer reaches of the 70-plus known Jovian system.

It was German mathematician, astronomer, and astrologer Johannes Kepler (1571–1630) who first suggested the existence of bodies orbiting Mars. Up until then, it was believed that the red planet was moonless. Kepler was obsessed with Mars and fascinated by the "Martian problem," wherein the planet appeared to go backwards across the sky before correcting itself and heading off in the right direction. Kepler boasted that he would account for the problem in eight days, with the actual resolution taking eight years! Despite this setback, the discovery was groundbreaking, as Kepler introduced his three defining laws of planetary motion:

1. Despite Copernicus concluding that all orbits are perfectly circular, they are not. Kepler stated that planetary orbits are elliptical and, in addition, moved the proposed position of the Sun from the center of the spherical Universe (as Copernicus believed) to one side of the orbits.
2. Planets do not move at a constant speed, but slow down further from the Sun.
3. The relationship between the orbits of any two planets is determined by their distances from the Sun, making their positions rigidly predictable (this solved the infamous Martian problem).

In the early 1600s, Kepler reasoned that if Earth had one moon and Jupiter had four moons, then Mars must have two moons. In 1726, Jonathan Swift (1667–1745) wrote of Mars' moons in *Gulliver's Travels*, even assigning orbital diameters and periods that incredibly were not far off the known quantities. Swift, an essayist, poet and cleric, produced substantial calculations to reinforce his belief that Martian moons existed, his later work probably combined with input from his close friend John Arbuthnot (1667–1735), a physician and mathematician. Writer Francois-Marie Arouet (1694–1778), who wrote under the nom de plume Voltaire, also mentioned these moons a few years later in his story *Micromegas*. In honor of the contributions of Swift and Voltaire, two craters on Deimos were subsequently named Swift and Voltaire.

The actual discovery of the Martian moons is attributed to American astronomer Asaph Hall (1829–1907) with the use of the Clark 26-inch refractor at the Naval Observatory in Washington. Hall sighted Deimos on August 12, 1877, following six days later with that of Phobos. At the time, Hall was actively searching for moons around Mars, with an earlier sighting being thwarted by bad weather, which meant he was unable to confirm what he had seen until a later date.

Phobos measures just below 23 km in diameter, with an orbital period of 7.66 hours. Deimos is slightly smaller at just under 13 km, with a far greater but still rapid orbit of 30.35 hours. Phobos orbits at a distance of 5,955 km from the Martian surface, while Deimos is much further away, at 20,069 km.

Phobos and Deimos bear more of a resemblance to asteroids than they do to our own Moon. To start with, both are small—among the smallest known moons in the Solar System. Secondly, both are made up of material akin to carbonaceous chondrites, which means they were likely formed from material much further away from the Sun than Mars. This would put them in the class of asteroids rather than planetary moons, adding credence to the Capture Theory. Their shape reinforces this notion, as both are elongated, as opposed to more rounded moons.

There is a problem with the Capture Theory for Phobos and Deimos, as there is with our Moon. In the case of the Martian moons, both take a stable, near-circular orbit around Mars—something captured bodies don't tend to portray. Thus, as with our own Moon, the other theories could apply, wherein the moons were spawned from the violent birth of Mars; the result of an impact; or perhaps they simply formed as the red planet did, as debris left over from the creation of Mars.

The problem with the Capture Theory in regard to our own Moon is figuring out what mechanism actually allowed for its capture itself. The theory would have to rely on a remarkable set of circumstances that allowed the

Moon to be captured and established in the orbit that it has today. At best, the odds favor an elliptical orbit, and at worst, an altered trajectory that would send the Moon off and out of the Solar System. Given the size of the Moon and perhaps the atmosphere around a primitive Earth, there may have been some natural aerobraking, but not enough to halt or deflect such a massive body. Thus, with a collision-based origin seems more likely.

Other theories propound that two moons once existed around Earth, one colliding with the other to create our Moon. Another theory argues that the Moon actually belonged to Venus.

Martian Ancestry

Is it possible that life as we know it on Earth came from Mars, or as some scientists suggest, from further afield?

Panspermia, from the Greek for "seeds everywhere," was first speculated about in the writings of Greek philosopher and scientist Anaxagoras (500–428 BCE). The hypothesis states that the "seeds" of life exist all over the universe, propagated throughout space from location to location, one planet to the next. One such carrier of these seeds could well be an asteroid or a comet. The theory goes that these migrating bodies contain the very building blocks of life, carrying the proteins necessary to establish the simplest of life forms. Discovered on comets and in meteorite fragments, these traces of amino acids and other organic compounds may well have formed in space, long before their descent through the atmosphere of a lifeless planet. This concept remains a theory, as proving panspermia would require a massive survey encompassing vast regions of space and requiring the in-depth exploration of other planets.

Several variations exist on the panspermia hypothesis:

1. Life began once on Mars and spread to Earth via Martian meteorites
2. Life originated independently on both Mars and Earth. Cross-colonization may have then occurred
3. Life began once, on Earth, and was then propagated to Mars, where it may well have established itself
4. Life originated on both Mars and Earth. However, despite an exchange of debris in the form of rocks and dust between the two, no transfer of viable organisms occurred
5. Life did not originate on either Mars or Earth, but somewhere else and, by a series of possible means, was brought to Earth

The obvious alternative to the panspermia idea is that life began on Earth and stayed on Earth, with no transfer to another planet.

The meteorite discovered in the Allan Hills in Antarctica on December 27, 1984 brought the possibility of life arriving from Mars into the media spotlight. The meteorite, named Allan Hills 84001, contained what appeared to be evidence of Martian bacteria. Of course, this hypothesis was quickly disproved.

Later meteorites thought to originate from the red planet itself were subject to close examination that in the end sided in favor of a Martian origin. How was this proven?

Whereas all previous meteorites have a high age associated with them, such rocks have a relatively young ages, around 1,300 and 165 million years old, compared to their counterparts that date as far back as 4,000 million years.

Their textures are different, being igneous (formed from melted rock), implying that the body from where they came has been active more recently, in complete contrast to an asteroid or our own Moon.

The ratios of the elements found within these meteorites are different to those of asteroids and those of the Moon. They show that it is highly likely that these meteorite finds came from one parent body—probably a planet-sized body. All asteroids are too small to remain hot for very long, whereas a larger body such as a planet can keep melting the rock for a longer period of time.

It was the gases trapped within such meteorites that confirmed the rocks born of Mars had made it to Earth. The makeup of these Martian meteorites included black glass, formed by shock melting and probably produced following the impact of another body on Mars, throwing this particular chunk of rock clear of the planet's surface and far into space. The air trapped within the black glass matches the composition of the air on the surface of Mars, which was measured by the American Viking lander probes when they visited the planet in 1976.

Martian meteorites have been collected from all around the world—from France, India, Egypt, America, Nigeria, and Brazil—confirming the wide global distribution of not just Martian meteorites but of all types of meteorites.

Classification of Martian meteorites falls into one of three groups, named after three of the best-known meteorites of this type: Shergottites, Nakhlites, and Chassignites, referred to as SNC meteorites.

Shergottites are the most common of the Martian meteorites, accounting for by far the largest proportion of finds. Named after the Shergotty meteorite that fell to Earth at Shergotty (now Sherghati) in the Gaya district of

India on August 25, 1865, this class contains three subgroups based on chemical content: basaltic, olivine-phyric, and lherzolitic.

The Shergotty meteorite is composed of solidified magma, produced during great volcanic activity on Mars some four billion years ago. It weighs 5 kg, with the meteorite's main mass a resident at the Museum of the Geological Society in Calcutta. it has a similar composition to terrestrial dolerites. Its fall to Earth was witnessed and logged at 09:00, with an accompanying description of several detonations being heard.

Nakhlites have a very different composition to Shergottites and are thought to be considerably older, their origin probably stemming from the impact of a large body onto the surface of Mars. They are named after Nakhla, a meteorite that fell in El-Nakhla, Alexandria, Egypt on June 28, 1911. The fall of the Nakhla meteorite, also (curiously) logged at 09:00 like the Shergotty fall, was witnessed by many of the local residents living in small settlements, who described its descent from the northwest as accompanied by a trail of white smoke and loud booms. Meteorite debris was collected from a wide area, with a total of 40 pieces gathered. The stones ranged in size from 20 g to 1,813 g, with an estimated total weight of 10 kg.

The Nakhla meteorite has been something of a headline maker. Apart from the fall itself, a legend has sprung forth about the Nakhla dog, which, according to a local farmer, was struck by a fragment of the rock and reduced to ashes. There remains doubt over the story and its credibility but it is nonetheless a great talking point.

More seriously, the Nakhla meteorite became the center of much controversy. In March 1999, examination of the meteorite using a scanning electron microscope (SEM) by a team from NASA's Johnson Space Center revealed small, rounded particles within the rock. The researchers, led by Dr. David McKay (1936–2013), suggested that the particles were in fact mineralized remnants of bacteria that once lived on Mars. Various amino acids were also discovered, but there was disagreement over their origin. In February 2006, further research conducted on a sample of the meteorite held by London's Natural History Museum found a carbon-rich substance filling cracks and channels within the rock. There was a remarkable resemblance between the substance found and the effects of bacteria observed in Earth rocks, thus discrediting the original theory.

Chassignites meteorites have similar formation ages to those of Nakhlites. Named after the Chassigny meteorite, which fell on Chassigny, Haute Marne, France, on October 3, 1815 at 08:00, one of its most striking composition differences is the presence of noble gas, probably indicating its source at the mantle of a larger body.

Returning Home?

It's much easier from rocks to travel from Mars to Earth than the other way around, with scientists continuing to speculate about whether life did hitch a ride from Mars to Earth in the form of a microbe. It would be a great irony if humankind were to journey to Mars for the first time, only to discover that in fact, they had simply returned home!

Chapter 11

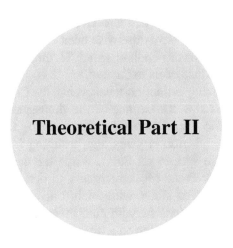

Theoretical Part II

Daring to Believe

When we consider that for every star in our Milky Way, there is believed to be one planet, it would appear that the possibilities of life on other worlds are boundless. Yet, we remain very much alone, which in itself is a very sobering thought.

In truth, we don't know everything—in fact, we know very little, and a substantial proportion of what we think we know is open to interpretation and debate. There are more dimensions, labyrinths, and quandaries than we could ever imagine, and we must be prepared for the possibility that as humankind we have made some severe errors of judgement, some woeful inaccuracies that have potentially blinded us from what the startling reality of all may well be. Such grand revolutions in perspective have and will continue to happen over the course of our scientific studies.

The fragility of our existence in the universe is apparent. There is still a deafening silence within our own neighborhood, but a silence that logic and statistics tell us must one day be broken. Is this inevitable?

In some fields of science, it does not hurt to speculate and make reasonable assumptions based on what we already know. Conjecture is after all part of the learning curve, using knowledge that we already know and intertwining it with that which remains uncertain. There are few of us, from our ancient ancestors to us modern astronomers, who haven't at one time set aside the instruments that we are using to calculate, estimate, and evaluate, to simply look at the night sky and ponder. The balance between fact and fiction is a necessary state, one that has led to countless proven discoveries in the field that would not have otherwise been considered.

We must be careful in our endless search for life, bracing ourselves for finding the possible cause of its extinction.

When conceptualizing extinction-level events, the Cretaceous–Paleogene extinction event that saw the demise of the dinosaurs most immediately springs to mind. The event, also referred to as the Cretaceous–Tertiary extinction, occurred 66 million years ago, resulting in the elimination of 75% of plant and life species on Earth. While the devastation was widespread, there were survivors. Most mammals, turtles, crocodiles, salamanders, and frogs survived, along with snails, bivalves, starfish, and sea urchins. Birds too escaped, along with plants that were able to adapt to the climate. This will to survive allowed a different chapter to begin on Earth, a molding and nurturing of new life out of extremes even as most other forms of life were terminated.

The survival of birds is of particular interest, as their survival—in part due to their adaption to live on the forest floor following the event—marks a turning point in Earth's biological history. The study of fossilized plants and ornithological data supports the theory that birds came to dominate the planet during the aftermath of this event. The global deforestation that followed the event would have accounted for the extinction of most flowering plants, destroying the habitats of tree-dwelling animals. It is believed that ground-dwelling bird ancestors, probably small and partridge-like in form, were able to adapt to life by surviving in a treeless world, making use of what limited offerings were available on the forest floor, eventually taking to the trees once more after the flora had sufficiently recovered. Once this happened, birds began to adapt to living again amongst the trees, acquiring shorter legs than their ground-dwelling ancestors and various specializations for perching on branches. The time frame for such a regeneration would have taken thousands of years. From here, as birds further evolved, great diversity took place, with ostriches and their relatives, chickens and their relatives, and ducks and their relatives all springing forth from the ancestral bird that once roamed the forest floor, millions of years ago.

Two main hypotheses still circulate regarding the Cretaceous extinction. The first and most touted is that the event was due to large impact from a rogue asteroid, perhaps even a comet. The other credits the extinction to a massive bout of volcanism. Whichever one sides with, the outcome was the same: a huge shift in climate where the skies were leaden from impact material from the asteroid strike, or volcanic material thrown up into the atmosphere from a series of long-lasting eruptions. With the Sun blotted out, greenhouse gases locked into the atmosphere would have caused the temperature to soar, and both plant life and animals would have struggled to survive in the dark years that followed.

The theory that the extinction of the dinosaurs occurred from an impact stems from the discovery of a particular layer of metal in rock dated precisely to the extinction event. This iridium-rich layer is found across the world, on land and in the seas. The presence of iridium is rare on Earth, but not so in meteorites. Scientists speculated that it derived from an impact by an extraterrestrial body, which scattered the debris on a global scale. The iridium formed a cloud that dispersed and spread around the world, eventually falling back to Earth and settling into this compact layer.

The theory is further supported by the discovery of a 180-km-wide crater carved out of Mexico's Yucatan Peninsula, called Chicxulub, since dated to 66 million years ago, which would coincide with the event.

However, iridium is also rich at the Earth's core. Thus, volcanism would also explain its presence. A substantial and sustained outpouring of magma from the Earth's core could have covered up to one million square miles of the Earth's surface, with a level of magma some 2.4 km deep. This could easily account for the iridium. A period of such volcanism has been dated to 65 million years ago, causing the same effect as an asteroid impact, with airborne dust blocking out the sunlight and starving the planet of natural light, directly damaging the ability of plants to photosynthesize and severely interrupting the food chain. In the gloom that followed, those that were not killed outright by the volcanic activity would have simply starved to death.

Tunguska Event

Much can be derived from a more recent encounter with a well-documented asteroid. On the morning of June 30, 1908, near the Podkamennaya Tunguska River in Siberia, a comet-like object blazed through the air. A brilliant, flaming fireball lit up the morning sky, exploding into showers of burning debris as it did so and leveling everything that stood before it. Believed to have been a superbolide, a large size of bolide or fireball meteor, the body disintegrated at high altitude, with remnants (including what was left of the main projectile) flattening an area of trees some 60 km across. The resulting prostrate wood pointed back to the direction of the impact site.

In total, it has been estimated that 80 million trees were felled in the taiga forest, scorched rather than burned, as the initial flames from the descending object lit them, while the trailing passage of wind quickly extinguished any flames.

The sonic boom shattered windows in the nearest town some 60 km distant, with residents in the town reporting a sudden blast of heat from the impact, some even being blown off their feet. There is evidence to suggest people were actually singed by the intense heat. The reports noted two human fatalities. One instance tells of a local deer herder who was thrown against a tree following the blast. Another report mentions an old man dying of shock. Both reports remain unsubstantiated. The loss of a single human life is tragic, and this occurred in a sparsely populated area. One is left to reflect on the potential devastation and loss of life were the superbolide to have fallen in a populated area. Many reindeer and other wildlife perished in the event.

For those around to witness the object's descent both near and far, a great, almost blinding flash was observed, with minor and much smaller debris from the object that didn't initially make it to the ground later burning up as meteors in the night sky. Right across Europe, over a sequence of nights, a nighttime glow was also observed. Seismic rumbles from the impact were felt as far away as the UK.

The blast from the Tunguska event has been calculated as measuring around one third as powerful as the largest and most powerful ever detonated bomb, the Tsar bomb. The Tsar bomb (King of Bombs) was detonated on October 30, 1961. The Chixculub asteroid that could have been responsible for the demise and termination of the 160-million-year reign of the dinosaurs generated a force one billion times greater than the atomic bomb dropped on Hiroshima at the close of the Second World War.

Chelyabinsk Meteor

The Chelyabinsk incident was a wakeup call for Earth, with a small but powerful body reminding us of the ever-present threat from above.

On February 15, 2013, a small asteroid broke up over the city of Chelyabinsk in Russia. The shockwave shattered glass and injured approximately 1,200 people. The blast briefly outshone the Sun and was stronger than a nuclear explosion, triggering detections from monitoring stations as far away as Antarctica.

The meteor measured approximately 17 meters in width and weighed a sizeable 10,000 tons. It grazed the Earth's atmosphere, shattering as it did so into many fragments that burned up with tremendous friction in a massive fireball. The estimated energy output release was 500 kilotons, approximately 30 times the yield of the Hiroshima bomb.

The origin of Chelyabinsk remains the object of speculation, with some scientists believing it to be a fragment of asteroid 2011 E040, which is known to have frequent encounters with Venus, the Earth, Moon, and Mars.

Studies determined that the Chelyabinsk impactor (the body responsible for the fragment that fell to Earth) most likely passed a gravitational keyhole (a portion of a planet's gravity that would alter the orbit of a passing body) on February 15, 1982, during a previous encounter with the Earth. At the time, the closest distance was approximately 224,000 km. As a result of that close encounter, the trajectory of the Chelyabinsk meteoroid was changed, providing it with the right conditions that drove the meteoroid towards the Earth three decades later.

By coincidence, the Chelyabinsk incident occurred at the same time that an asteroid was flying close to Earth at a distance of 27,000 km. It was named 2012 DA14, with an estimated size of 40 meters. NASA pointed out that there was no connection between the two, as the Chelyabinsk body was traveling in the opposite direction to 2012 DA14.

In the days following the explosion, the area became swamped with meteorite hunters from across the world, keen to track down pieces of the impactor, which had exploded high up in the atmosphere above Russia and therefore made for a wide field of debris distribution. On February 18, 2013, the first reports emerged that pieces had been located around Lake Cherbarkul, 70 km north of Chelyabinsk. At the lake, which was covered in a layer of ice, it was also discovered that something substantial had penetrated its surface, sinking down to the depths below. After searching the lake bed 20 meters below (which strangely enough was later found to harbor possible extraterrestrial rocks buried in the mud sediment), the hunters recovered a 1.5-meter rock, which duly split into three segments live on Russian television.

Together, the three stones weighed in at 570 kg. The meteorites sported all the signs of a trip through the Earth's atmosphere, including a fusion crust—a shiny glass layer of black material that forms when the outer portions of the rock melt. The rocks also appeared to have regmaglypts, shallow surface indentations that appear like thumbprints.

On the same day as the Chelyabinsk incident, the US House of Representative's Science, Space, and Technology Committee said that it would hold a hearing to discuss asteroid threats to Earth and how to mitigate them on top of NASA's current efforts.

2018 GE3

Despite many astronomical eyes being trained on the heavens above, stark warnings about potential last minute collisions are issued on regular basis.

On April 15, 2018, and with virtually no warning, a "Tunguska-Class" asteroid about the size of a football field flew across the Earth–Moon system. Tagged 2018 GE3, the asteroid was spotted just one day before it plunged towards territory near the Earth. At around 48 to 110 meters in diameter, it was thought to be five times bigger than the meteor that exploded over Chelyabinsk and approximately 3.6 times the size of the object that flattened 2,000 km²of Siberian forest over Tunguska in 1908.

First sighted in the early hours of that Saturday morning by the Catalina Sky Survey (a NASA-funded project supported by the Near-Earth Object Observation Program (NEOO)) and Steward Observatory in Arizona, 2018 GE3 sped by just 24 hours later, coming within about half of the distance between the Earth and the Moon—some 192,000 km—at a speed of 106,000 km/h.

Observers argue that 2018 GE could have been picked up as early as March 30, 2018. To be fair, survey telescopes need to be pointing in the right direction to spot anything of this nature and, if they are not, such objects will be missed. It is not uncommon for asteroids to be picked up at this close distance. This particular incident made 2018 GE the 32nd known asteroid to fly by Earth within one lunar distance (LD) since the start of 2018. But whether they are sighted millions of miles out or several thousand miles away, no firm plan is in place to deal with the threat, should it be confirmed that the body will collide with the Earth.

2018 GE3 is an Apollo asteroid, hailing from a group of asteroids that cross the orbit of the Earth. The first of the Apollo asteroid family to be discovered was 1862 Apollo, found by German astronomer Karl Reinmuth (1892–1979) on April 24, 1932. Apollo 1862 is a PHA (potentially hazardous asteroid). The Chelyabinsk meteor was also member of this family. This group forms the largest of near-Earth objects, with a collective number in the region of 10,000. 2018 GE3 was most likely sent in our direction as a result of a collision with another family member.

Fossil dating emphasizes that impact incidents on Earth are not lone or rogue occurrences, but in fact are just one of perhaps as many as nine or more episodes that the Earth has suffered in its turbulent history. All of these events have taken place in the last half billion years. Based on that scenario, it might be logical to think that it is only a matter of time before another extinction-level collision occurs. Many argue that the Earth and Solar System's past was more volatile, and therefore the planet was more susceptible to such occurrences. In this state of flux, more debris and fragments were flying around, perhaps leading to a greater likelihood for impact. Possibly so, but dwell on this: there may well be a lot less debris around and, as science advances, the ability to spot and track it now gives us an edge. Yet whether we spot it or not, we are in truth utterly defenseless at this point in time. All of our tracking only allows us to see the face of the intruder before the act is committed, and nothing more.

Might it be possible that extinction-level events have nothing whatsoever to do with rogue comets or asteroids? Could such global devastation actually have been caused by a massive gamma-ray burst?

These huge galaxy-wide bursts of electromagnetic radiation are the result of the dying explosion of a hypernova, or when two neutron stars collide. Hypernovae seem to be the more likely of the two, with a violent explosion far more aggressive than that of supernovae. Considered to be one of the most devastating events in the universe, a hypernova would certainly account for a widespread annihilation of whatever life existed on planets within its reach. Lasting anywhere from a few milliseconds to several minutes, one such hypernova has the power within its gamma-ray burst to wipe out life across huge swathes of the galaxy—a sort of cosmic clearing and leveling of the playing field.

Research has determined that the probability of the Earth having been on the receiving end of one of these destructive gamma outbursts is 60%, given the time frame of a billion years. That statistic rises to 90% when extending the parameters to the last five billion years. With this information, it would seem more than just a speculative guess that a gamma-ray burst was responsible for the first great extinction event that took place in Earth's history over 440 million years ago.

The Fermi Paradox

Italian physicist Enrico Fermi (1901–1954) together with American physicist and colleague Michael H. Hart (b.1932) posed the question as to why, given the vastness of space and the possible evolution of life elsewhere, we haven't actually encountered any alien life forms. Fermi, after whom one of the two kinds of particles in the universe was named (the "fermion," with the other being the "boson"), theorized that surely, given certain evolutionary circumstances throughout the universe, someone from somewhere should have made contact with us by now!

Famous for creating the first nuclear reactor, Fermi proposed that any civilization with a modest amount of rocket technology and the accompanying desire to use that technology to explore its surroundings in space could literally, if unopposed, colonize the galaxy. Fermi further proposed that, given the length of time at the disposal of said "conquerers," a vast swathe of space should already be under the control of one civilization, or a combination of several, all depending on how the expansion plans pan out.

According to some sources, Fermi never actually published a word on the subject of extraterrestrials. Much of what we know about his beliefs came through physicist Eric Jones, who collected three written accounts from the three surviving people present at the luncheon: Emil Konopinski, Edward Teller, and Herbert York. They recounted that while musing over a cartoon in *The New Yorker*, which depicted aliens disembarking from a saucer, Fermi is alleged to have said, "Where is everybody?" As the chatter continued, the general consensus at the time of the luncheon was that Fermi was questioning the feasibility of interstellar travel, not that of the possible existence of extraterrestrial aliens. As with all accounts made after the fact, it remains possible that Fermi's quote is misrepresented.

Michael H. Hart wrote an article entitled "An Explanation for the Absence of Extraterrestrials on Earth" in the *Royal Astronomical Society (RAS) Quarterly Journal* in 1975. Hart was possibly one of the first scientists to properly explore Fermi's paradox in any detail. He outlined four arguments:

1. Aliens never came because of a physical difficulty that could be related to astronomy, biology, or engineering
2. Aliens chose never to come to Earth
3. Advanced civilizations arose too recently for aliens to reach us
4. Aliens have visited in the past, but we have not observed them

Fermi's classic statement "Where is everybody?" resonates with many who believe contact in some shape or form should have long been established. With as many stars as there are in our galaxy—upwards of 400 billion—there are roughly an equal number of galaxies in the observable Universe. In other words, for every star in our own Milky Way, there is a galaxy out in the universe to match it number-wise.

However, the very absence of evidence is not evidence of absence.

If potentially habitable Earth-like planets orbited even 1% of the total stars in the universe, that would mean there could be upwards of 100 trillion Earth-like planets orbiting out there in space.

In 1980, mathematical physicist and cosmologist Frank J. Tipler, working out of Tulane University in New Orleans, Louisiana, contributed to the Fermi Paradox debate with his paper "Extraterrestrial intelligent beings do not exist" which was also published in the *Royal Astronomical Society Quarterly Journal*. Tipler's paper argued that since such beings aren't already on Earth, we are likely to be the only intelligence out there.

One must consider the level of technological advancement achieved by another potential civilization, and whether such developments make them capable of interstellar communications. Simple radio waves are transmitted across the Earth's globe on a daily basis, many continuing into space. SETI has been scanning the skies for decades in search of such a signal, but nothing has ever been received—that is, nothing that hasn't been disputed with a counter claim about what has actually been heard. Even if a sound is not challenged, the origin and source don't guarantee communication from an extraterrestrial lifeform—the sound merely remains unexplained.

The silence becomes more perplexing when one thinks about all the Sun-like stars that are in a more advanced stage of their life compared to ours, meaning that many civilizations could have already evolved while the Earth was in its infancy.

Kardashev Scale

Created by Russian astronomer and astrophysicist Nikolai Kardashev, the Kardashev scale is a method of measuring a civilization's level of technological advancement, based on the amount of energy a civilization is able to use for communication. The scale helps to group intelligent civilizations into three broad categories:

Type I: Civilization is able to use all of the energy on its planet. Also known as a planetary civilization, a Type I civilization has harnessed the ability to use and store energy that reaches the planet from a host star.

Type II: Civilization can harness all of the energy from its host star. If this looks remarkably like a Type I civilization, one has to imagine a Dyson sphere, a hypothetical structure that allows for the capturing of a great percentage if not all of the host star's total energy output.

Type III: Civilization has achieved the Type II civilization state but grown in its capabilities to enhance energy so much so that it now commands power comparable to that of an entire Milky Way galaxy.

Since the three were formulated, other proposed levels of civilization have also been put forward to compliment Types I to III.

One argument explaining the lack of signs for Type II or Type III civilizations simply states that no such civilizations exist. Possibly, everything is on the same level as the civilization on Earth—a Type I.

So, what classification do we fit into? In 1973, based on the energy use at that time, astronomer and cosmologist Carl Sagan estimated that the Earth represented a Type 0.7 civilization, not even a Type I. Even after factoring in more recent advancements over the decades, we still sit at around 0.73 or so. This is somewhat discouraging after 4.5 million years, but not when one considers what goes into actually achieving a Type I Civilization status.

At the very basic level (Type 0.1), we would be proto-humans, using sticks and other simple and basic tools to hunt and forage for food, without much clothing of any sort. As an individual within this society, one would need to defend oneself appropriately in order to earn mating rights, with dominance over the tribe governed by a hierarchical structure, whereby the fittest and strongest would attain the loftiest position. This status would in time be challenged by younger members, or when said self-imposed leader is either killed or eaten by some sort of animal. This lowest type of civilization would rely solely on natural resources, with fires only created by the intervention of lightning.

A Type 0.2 civilization follows, clothing used to provide warmth against the climate and the eventual harnessing of fire to light the night and roast the carcass of any hunted animal. Such a progression begins the path towards an overall incorporation of not just natural forms of energy, but also the manipulation of natural resources for more diverse usage. A 0.2 civilization would also incorporate the use of animals as transport vessels. Communication by a primitive form of smoke signals would also evolve during a Type 0.2, allowing for interaction with other tribes in other areas as a general expansion across the globe takes place.

The advancement would take the civilization through the Stone and Metal Ages, introducing better fuels such as coal and oil, with communities steadily growing in culture and societal standing. With the construction of large structures that harness water and wind power, it has reached the marker of a Type 0.4 civilization.

The Industrial Revolution follows with the widespread use of fossil fuels. Steam and electricity accelerate the rate of progress on a global scale, along with the advancement of communication transporting, encouraging trade with countries far afield.

From fossil fuels to nuclear fuel, and the correct and proper harnessing of this energy for the good of the planet, the civilization would eventually become a Type 1 Civilization. This is a society that can promote a united global community capable of harnessing all the solar energy that reaches the Earth, able to manipulate planetary weather systems and pull energy from other sources in the cosmos. Failure to evolve to this point and beyond will ultimately result in the total annihilation of all inhabitants on the planet, driving the evolution down to the days of sticks and foraging for food, or even further back.

The Great Filter

Another theory purporting that we really could be on our own is something called the Great Filter, proposed by American economics professor Robin Hanson in 1998. The Great Filter theory states that, at some point between pre-life to Type III intelligence, there is a "wall" that all or nearly all attempts at life come up against. There is some point in the evolutionary process that makes it extremely difficult, if not impossible, to progress beyond: this is the Great Filter. The Filter, which applies not just to our advancement but also to the progression of any Type III intelligence, appears to be a necessary rite of passage before intelligent life can manage to spread beyond its original planet and onto other worlds.

The Great Filter would account for the equally Great Silence that we continue to encounter, and very neatly both explains and realizes the staggering importance of humankind's place in the universe. This in itself begs three further questions. Is humankind one of just a few select civilizations to make it beyond the Great Filter, meaning that we are not alone, but the chances of another such civilization surviving in our locale are very remote? Are we the first to make it past the Great Filter, which would make us truly unique but also truly alone? Last but not least, have we yet to encounter the Great Filter, which still lies ahead of us and which we most likely will not be able to surpass? Some scientists tout that climate change is the Earth's very own Great Filter.

There is also the prospect that in fact, there is more than one Great Filter. Hanson defines his Great Filter theory as "the sum total of all of the obstacles that stand in the way of a simple dead planet (or similar sized material) proceeding to give rise to a cosmologically visible civilization."

Was the extinction of the dinosaurs by whatever cause a derailment in the process of the Great Filter, meaning that we had to start all over again? Was the extinction a freak episode that sent life on Earth on an entirely different path from what would have occurred had the dinosaurs not been wiped out? What if the most famous die-off that ended their reign some 66 million years ago had not occurred?

There have indeed been many trials and tribulations to pass through in order to get to a stage where we can consider our own place in the universe, something scientists refer to as the "observation selection effect." This effect suggests that anyone who is pondering their own rarity in the universe is inherently part of an intelligent life and, in analyzing their own place, the thoughts they ponder and conclusions they draw are identical.

The Fermi Paradox, the Great Filter, and the Great Silence all have to afford adequate room for the Rare Earth Hypothesis (REH). This is the suggestion that the emergence of life was extremely improbable and that humankind is a one-off. What has developed on Earth is a set of circumstances that is quite remarkable and unique, with many random factors converging in a perfect storm to create a situation that couldn't possibly be duplicated anywhere else in the universe.

The REH was first proposed by geologist Peter Ward and astrobiologist Donald E. Brownlee and draws on a much wider set of random events and happenings to suggest how life began on Earth. This takes into account not just the Earth's orbital position with regard to our Sun, atmospheric conditions, or the presence of necessary organisms, but a much wider series of conditions, including the positioning of our own Moon, the position of the planets that make up our Solar System and their own unique compositions, and the positions of the stars.

Other Extinction Events

In these discussions, we must not forget to consider not only how foreign bodies may have affected the Earth, but also Mother Nature's own turbulent past, and her ability to be just as disruptive to her current custodians as she has been in our planet's long history. Take for example the Permian–Triassic extinction event, also known as the Great Dying, which took place around 252 million years ago.

The Permian–Triassic extinction event was one of the most severe mass extinctions in Earth's history, occurring long before dinosaurs and mammals evolved. Geologists and archaeologists know that at the end of the Permian period, something was responsible for the demise of up to 90% of the planet's species in the sea and on the land. Less than 5% of the animal species in the seas survived, with less than a third of the large animal species on land able to exist beyond the destruction.

While initially, an asteroid collision with Earth was an obvious candidate to explain the Permian–Triassic extinction event, there were many aspects that did not conform to the usual evidence trail associated with an impact event. Rocks dating from the age of the dinosaur extinction contain traces of asteroid that when analyzed and correlated with widespread catastrophic events, almost irrefutably lay the blame at the feet of a substantial chunk of cosmic debris.

However, with the Permian–Triassic extinction event, there are anomalies that cast a wider net regarding the potential culprit. Sediments containing fossils from the end of the Permian are rare and often inaccessible, although one site exists that still harbors the victims of the event, a scrubland area in South Africa known as the Karoo. Here, rocks dating from the period show the end of the dominance of one particular group of mammal-like reptiles known as the synapsids, the first great dynasty of land vertebrates. The synapsid fossils indicate that at the transition period between Permian and Triassic, something of devastating proportions occurred.

Aside from the usual asteroid strike being touted as the likeliest offender, other theories were proposed to identify the trigger for the Permian–Triassic extinction event, including a "killer" from the depths of the sea. This intriguing notion stems from the fact that scientists know that the deep ocean lacked oxygen in the late Permian period. Researchers suggest that rocks that formed under shallow water at the time of the extinction suffered from anoxia, or oxygen depletion. We know that pollution sometimes turns water anoxic, especially in regions that lack good circulation, resulting in the dying off of marine life. It has been proposed that during Permian times, the entire ocean may have stagnated, perhaps due to a lack of ice caps, which along with the subsequent temperature differences between polar and equatorial regions created convective currents. Polluted water, perhaps also poisoned with CO_2, may therefore have accounted for the demise of marine life. But even with rising seas levels, would it have accounted for the virtual annihilation of other life?

One of the most plausible explanations for the Permian–Triassic extinction event is a series of massive volcanic eruptions. Another mass extinction event took place before the Permian–Triassic, roughly 260 million years ago at the end of the Capitanian age, which accounted for the demise of up to 80% of all land animals. Evidence suggests that while the Permian event was mainly sea-based, most of the land animals had already gone extinct during this earlier Capitanian event. Indeed, the Capitanian event may well have ushered in the Permian event because of the damage left on ecosystems depleted of certain species.

And what of the Capitanian event itself, one that dwarfs by comparison the Permian event? Further examination of other fossils uncovered from the Karoo scrubland suggests that the formation of a mountain range in South Africa was one of the drivers behind the event. The teeth of one of the fossils, that of a plant-eating, mammal-like reptile, the *Diictodon feliceps*, contained evidence that the Karoo area dried out during the time of the extinction, coinciding with the emergence of South Africa's Cape Fold mountains. The birth of the mountains would have prevented moisture from the sea from reaching the interior of the land, which caused the area to dry up and in turn led to the death of its inhabitants.

The Greenhouse Effect

We only have to turn our attention to our nearest planetary neighbor to see the devastating effects of a runaway greenhouse effect environment. As dazzling as Venus may be in the morning or evening sky, it is that famous shroud of toxic cloud that reminds us of what happens when a climate changes beyond repair.

The term "greenhouse effect" was first coined by scientist Andrew Ingersoll in a paper that described the atmosphere of Venus.

We know that greenhouse gases on Earth are rising at a much greater rate than the time that saw the demise of the dinosaurs, and possibly farther back in history. Carbon dioxide is being added to the atmosphere at least ten times faster than during the last major warming event around 50 million years ago. Increasing levels of CO2 will also lead to temperature rises and ocean acidification at a rate that Earth's delicate ecosystem may, like Venus, struggle to cope with.

Rapid increases in greenhouse gases and mass extinctions are correlated, with several abrupt warming events occurring 56 million and 52 million years ago, the most prominent of these being the Paleocene Eocene Thermal Maximum (PETM). This event resulted in the extinctions of life in the deep ocean, with atmospheric temperatures thought to have increased by 5 to 8 °C (41 °F to 46 °F) within just a few thousand years.

Some scientific quarters conclude that because of human overpopulation and overconsumption and the resulting loss of literally billions of animals over previous decades, we are already in the midst of a "sixth mass extinction." Otherwise referred to as the Holocene or Anthropocene extinction, the sixth mass extinction refers to the eradication of a vast number of species during the Holocene Epoch (the current geological epoch), mainly as a result of human activity. The extinction includes the loss of many forms of plant life, with mammals, birds, amphibians, reptiles, and arthropod families also succumbing to such a fate.

The five major extinctions events in Earth's history are:

1. Ordovician–Silurian extinction event
2. Late Devonian extinction event
3. Permian–Triassic extinction event
4. Triassic–Jurassic extinction event
5. Cretaceous–Paleogene extinction event

The Higgs Boson Particle

One of the great byproducts of open thinking is the casting of a wide net, which taps into many areas that might otherwise not be considered. Such an avenue for investigation involves the Higgs boson particle.

The Higgs boson was named after two people: British theoretical physicist Peter Ware Higgs (b.1929) who, along with colleagues, first proposed the existence of the particle back in the 1960s, as well as mathematician and physicist Satyendra Nath Bose (1894–1974), a pioneering figure in the early days of particle physics. Bose most notably worked with Albert Einstein, developing a theory regarding the gas-like qualities of electromagnetic radiation.

In order to understand what the Higgs boson is, we first have to place the particle in the context of the cosmos as a whole. We therefore turn to the most subscribed to model, called the "Standard Model."

Initial findings brought the discovery of atoms, then protons, neutrons, and electrons, and finally quarks and leptons. Aside from matter, the universe also contains forces that act upon that matter. The Standard Model of the cosmos has given us more insight into the types of matter and forces that are at work than perhaps any other theory has allowed.

Developed in the early 1970s, the Standard Model dictates that the universe is made up of 12 different matter particles and four forces. Among those 12 particles are six quarks and six leptons. Quarks make up protons and neutrons, while members of the lepton family include the electron and the electron neutrino, its neutrally charged counterpart. Physicists believe that leptons and quarks cannot be broken into smaller particles. Along with the 12 particles, the Standard Model has four forces: gravity, electromagnetic, strong, and weak.

The Standard Model's composition has been very effective bar its failure to fit in gravity, but despite this, physicists have pushed the boundaries of understanding beyond the 12 particles, suggesting the existence of certain other particles years before such particles could be empirically verified. For a long while, one piece of the jigsaw remained elusive—that of the Higgs boson. Physicists believe that each of the four forces has a corresponding carrier particle, or boson, that acts upon matter. This in itself is a difficult concept to grasp, as we tend to believe that the four forces are not tangible, "hands-on" pieces of solid reality; in truth, though, the forces are as real as matter itself. One could say that the bosons themselves are a kind of weight anchored by mysterious "rubber bands" to the matter particles that generate them. This analogy would make them a more tangible entity to visualize, and with that visualization in mind, one can think of the particles constantly snapping out of existence in an instant, or equally being capable of intertwining with other rubber bands attached to other bosons, imparting force in the process.

With the four forces each owning specific bosons, physicists believe that the Higgs boson might have a similar function in transferring matter itself. The Standard Model does not allow for matter to have mass without the Higgs boson. Physicists suggest that if all particles have no inherent mass, perhaps they could gain mass by passing through a field. This field, known as the Higgs field, could affect different particles in different ways. If the Higgs boson exists, then everything that has a mass receives it by interacting with the Higgs field, which occupies the entire Universe. The Higgs field would require a carrier particle to affect other particles, and that particle is known as the Higgs boson.

On July 4, 2012, the ATLAS and CMS experiments at CERN's Large Hadron Collider (LHC) announced that they had found a particle that behaved the way they expected a Higgs boson to behave, the so-called (and much controversially so) "God Particle." ATLAS (A Toroidal LHC Apparatus) is one of the seven particle detector experiments constructed at the LHC, along with CMS (Compact Muon Solenoid). The ATLAS experiment observes phenomena associated with highly massive particles, taking advantage of the LHC's ability to generate unprecedented energy levels, while the CMS experiment, along with searching for the Higgs boson, seeks out extra dimensions and particles that could make up dark matter.

The particle shows "Higgs-like" qualities, but with a significant degree of uncertainty—perhaps not the Higgs boson particle itself. The "God Particle" description was coined by physicist Leon Max Lederman (b.1922) and was the title of his book that discussed the Higgs boson particle. Lederman insists that his publisher rejected the original idea of the "Goddamn Particle." The term has muddied the waters around the particle itself, and one needs to make a clear distinction between this label and the actual theorized workings of the Higgs boson itself.

A chance exists that should one of these particles collapse in a distant corner of the universe, a bubble of expanding vacuum energy could be produced that might eventually expand to a state where it encompasses everything. In essence, we would encounter a massive wall of negative energy engulfing and simultaneously ending everything in its path—a void that would gobble up the universe. Thus, could it be that the discovery of the Higgs boson could also be the universe's ultimate downfall? The very particle that was being sought ultimately leading to that which finally dictates the end?

As we know, the Standard Model doesn't yet account for gravity, and there are also blind spots in the theory when addressing dark matter and dark energy. Some have suggested that Higgs stabilizers may exist, allowing for the Higgs boson to become more stable and less likely to collapse. Given the length of time that the universe is already thought to have existed, perhaps these stabilizers are already in place and have been so since the very beginning.

Multiverse Theory

The multiverse theory states the existence of a set of various possible universes including our own—universes within the multiverse called "parallel universes," "other universes," or "alternative universes." These parallel universes can be broken down into levels:

Level 1: The rules of probability state that somewhere in this vast universe, another Earth exists. Given the high potential for a number of other Earths, the events that have played out here would be played out in an identical fashion elsewhere. Somewhere, someone is going to be exactly the same as you are now, in a world where all the events that are taking place are exactly the same as everything happening at this moment in time. Assuming that the universe is finite, and that every single possible configuration of particles in a Hubble volume or Hubble Sphere (a spherical region of the observable Universe surrounding an observer beyond which objects recede from that observer at a rate greater than the speed of light due to the expansion of the universe) takes place multiple times, reaching a parallel universe would be virtually impossible, ruling out the possibility of meeting oneself. In any case, the biggest stumbling block would be exactly where to start looking for a Level 1 parallel universe.

Level 2: In this universe, regions of space are still undergoing an inflation phase, continuing to expand following the Big Bang. If this is the case, then the regions of space in a state of inflation are getting further and further away from us, literally expanding at the speed of light, giving us no hope of reaching them.

Level 3: This is the consequence of a many worlds interpretation (MWI) from quantum physics, in which every single quantum possibility inherent in the quantum wave function becomes a real possibility in some reality. What sets a Level 3 universe apart from other levels is that this universe exists in the same space and time as our own universe, but despite this, there remains no way to access one another. Direct contact is impossible, but we are constantly interacting with a Level 3 universe, as whatever you do at one particular second causes a split of your "now" self into a finite number of future selves, all of which are unaware of any of the other's existence.

Level 4: Here, any universe that can be arrived at through the *mathematical democracy principle* could exist. In other words, any universe that physicists declare as mathematically possible has an equal possibility of actually existing.

The multiverse theory, or meta-universe, also states that the universe will cease to exist in around five billion years' time, ironically around the same time our own Sun runs out of its hydrogen and helium fuel, its core contracting with the expansion of cooler, less bright outer layers in the formation of a red giant.

Having already existed for an estimated 14 billion years, our universe according to the multiverse idea is but one of an infinite number of universes, an all-encompassing structure of an infinite number of universes, each of which are capable of spawning an infinite number of daughter universes. The prediction that our universe has five billion years left stems from the theory of eternal inflation, which states that our universe is part of a multiverse.

Eternal inflation is an extension of the base theory of inflation, which answered many of the questions posed against the workings of the original Big Bang theory. According to the early models of the Big Bang, groups of matter that are now on opposite sides of the universe are too far apart to have ever been in contact with each other, meaning that the early universe should have been clumpy in nature. Furthermore, at the rate that our universe is expanding, the overall shape should have curved over time, with the initial moment of creation having filled the universe with heavy, stable particles called magnetic monopoles. However, observations of radiation left over from the Big Bang contradict the early models, stating that the early universe was uniform rather than clumpy, that the shape of the current universe is flat and not curved. Further, magnetic monopoles have never conclusively been observed. That said, another theory does account for all of this. The Standard Inflation theory dictates that the universe experienced a period of extremely rapid expansion in its first few moments, eventually leveling off to create the flat, uniform universe that we know at present.

The potential existence of a Level 2 universe is supported by the theory of eternal inflation, along with the Ekpyrotic theory. The findings of eternal inflation mean that when inflation starts, it produces not just one universe, but an infinite number of universes.

The Ekpyrotic Model goes directly against the Big Bang theory. Whereas the Big Bang theory suggests a universe that began as very hot and intense, it actually began as extremely cold and nearly vacuous. The hot and expanding universe as we understand it today was the result of a collision that brought it up to a large but finite temperature and density. The Ekpyrotic theory also challenges the fundamental catalyst for the making of stars and galaxies, stating that there was no natural mechanism in place to create them, or indeed the larger structures that make up the universe. From there, the Ekpyrotic theory tends to meld with how the Big Bang developed. Its own explanation concerning the stars and galaxies involves the collision of two three-dimensional worlds moving along a hidden, extra fourth spatial dimension.

Behind the Ekpyrotic Model is American theoretical physicist and cosmologist Paul Steinhardt (b.1952) of Princeton University, New Jersey. The concept is considered a precursor to the more powerful and ultimately more convincing Cyclic Model of the universe. The term "ekpyrotic" derives from the Greek word "ekpyrosis," meaning "conflagration," referring to an ancient Stoic cosmological model that stated that the universe was created in a sudden burst of fire, not unlike the basis of the Ekpyrotic Model.

The Big Crunch

The Big Crunch is one of the scenarios predicted by scientists, in which the universe may end. Based on Einstein's Theory of General Relativity, the Big Crunch is the Big Bang in reserve, ending as it begun. Based on the premise that the universe's expansion due to the Big Bang cannot be sustained indefinitely, scientists argue that it will eventually stop and duly collapse in on itself, pulling everything in existence to a point where the largest ever black hole forms, consuming all in a Big Crunch.

Heat Death, Big Freeze, Big Chill

Heat Death of the universe, also known as the "Big Freeze: or "Big Chill," is the scenario resulting from the endgame of the second law of thermodynamics, which states that within a closed system, entropy can only increase. It infers that energy can only move from being more concentrated, more organized, and more useful, to being less concentrated, less organized, and less useful.

Let us take an individual example. As stars burn, they take usable, organized energy and convert that energy into its distributed, most unusable form. When a star burns out, the energy that it sent out before burning up still exists somewhere. All of the energy from all of those billions of dead stars is still out there— all the heat still exists, dispersed and unusable. It would require more energy to collect the used energy than that used energy would then be able to supply. Heat Death is the unavoidable and complete powering down of the universe in a way that cannot be reversed. A final state exists where there is no free thermodynamic energy or thermodynamic equilibrium, leading to the end of all activity in the universe.

The Big Rip

While a proportion of physicists believe the universe will gradually get colder and simply fade out of existence some 2.8 billion years from the present day, others believe dark energy could cause the universe to simply rip apart.

The Big Rip or sequence of rips would occur as the universe expands and constituents of matter within the universe start to separate from one another. The Big Rip relies on the assumption that the universe will continue to expand at a faster and faster rate, eventually causing it to literally tear or rip. The expansion would cause galaxies to separate, then planets, and eventually individual atoms. This gradual "ripping" would leave the universe entirely devoid of structure. However, the Big Rip is reliant on dark energy being a particular type, called "phantom energy." Phantom energy is present when the ratio of dark matter pressure to energy density is less than −1, which means the dark energy pressure is greater than its energy density. That in turn allows the universe to keep expanding outward until the universe is ripped apart. The problem in determining the likelihood of the Big Rip is the unknown ratio between dark energy pressure and its own energy density.

Oumuamua: The Mystery Visitor

On October 19, 2017, the Pan-STARSS 1 telescope in Hawaii spotted something quite bizarre gliding through our Solar System. The telescope, funded by NASA's Near-Earth Object Observations (NEOO) program, finds and tracks asteroids and comets nearby.

First observed from outside our Solar System, the object was described by researchers who first glimpsed it as an "oddball." The usual suspects of a comet or an asteroid were first touted to describe that rapidly moving point of light. It was soon found that the object had not originated in our Solar System—that it was an outsider, venturing into our neighborhood from interstellar space.

Having realized that all was not what it seems, available telescopes swung their gaze in the direction of the visitor, collectively focusing on the object for the next three nights in order to determine exactly what we were dealing with before it zoomed out of sight.

The object executed a slingshot maneuver around the Sun on September 9, 2017 at a breakneck speed of 315,000 km/h. Over the coming days, it revealed itself as a rocky, cigar-shaped object with a somewhat reddish hue, the result of millions of years of exposure to cosmic rays. The hue is similar to that of objects found in the Kuiper Belt, in the outer part of our Solar System.

The Kuiper Belt is similar to the asteroid belt but far larger, as much as 20 times as wide and 20 to 200 times as massive. Named after Dutch-American astronomer and planetary scientist Gerard Peter Kuiper (1905–1973), the belt is an elliptical plane in space spanning from 30 to 50 times Earth's distance from the Sun, or 4.5 to 7.4 billion km. Kuiper, who also discovered two natural satellites of planets—Uranus' satellite Miranda and Neptune's satellite Nereid—did not actually predict the belt's existence; that accolade came following the 1992 discovery of Albion, the first recognized Kuiper Belt object (KBO). Its full title is 15760 Albion, and it was the first trans-Neptunian object to be discovered after Pluto and its satellite Charon.

Albion was discovered by English astronomer David C. Jewitt and Vietnamese-American astronomer Dr Jane X. Lu'u using the facilities at Mauna Kea Observatory. Over the years, their finding revealed many more objects, with estimates citing potentially thousands of bodies more than 100 km in diameter and trillions of smaller objects, all traveling around the Sun in the Kuiper Belt. Pluto lies in the belt as well. Although unrecognized as being so at the time of its discovery, it was the first true Kuiper Belt Object.

Measuring 400 meters long and highly elongated—perhaps 10 times as long as it is wide—the cigar-shaped intruder was classed as an interstellar asteroid and officially designated the tag A/2017 UI by the International Astronomical Union, which created the category after its discovery. The object has another official name designated by its discoverers: Oumuamua (pronounced MOO-uh-MOO-uh), coined partly because of the location of the telescope that discovered it, and loosely meaning "a messenger that reaches out from the distant past." Other translations of the name include "a messenger from afar arriving first" and "scout."

Oumuamua's shape is surprising enough, not conforming to the more uniform stereotypical look of an asteroid. Observations suggest that the object had been wandering through the Milky Way for potentially hundreds of millions of years, and it was only by chance that it flew through our neck of the woods. This maverick is not attached to any star system; it is a true wanderer from deep space. Speculations about such objects had been debated for decades; with the appearance of Oumuamua came the realization that they really do exist.

An attempt during Oumuamua's fleeting appearance was made to gather as much data as possible, combining information from ESO's Very Large Telescope in Chile with other large telescopes. The accumulated results revealed that Oumuamua fluctuates greatly in brightness, classed as "dark" (absorbing 96% of the light that hits it) but varying by up to a factor of 10 as it spins on its axis once every 7.3 hours. No known comet or asteroid in our Solar System varies as widely in brightness, with such a large ratio(1:10) between length and width. Before Oumuamua's appearance, the most elongated objects known were no more than three times longer than they were wide.

Oumuamua seemed to be completely inert, without the faintest hint of dust encircling the body. It is thought to be composed of rock and perhaps metals, but no water or ice. Further speculation suggests that Oumuamua could contain an abundance of hydrocarbons, the building blocks for life that many scientists believe first arrived on Earth from asteroids.

Where Did Oumuamua Come from?

As Oumuamua ventured further away from the Earth, several large telescopes continued to track this mysterious visitor, namely NASA's Hubble and Spitzer telescopes. The object's outbound path took it about 20° above the plane of the planets that orbit our Sun. On November 1, 2017, Oumuamua passed the orbit of Mars, subsequently passing the orbit of Jupiter in May 2018. It is expected to push beyond the orbit of Saturn in January 2019.

The projected route out of the Solar System will take the object in the general direction of the constellation Pegasus. Given Oumuamua's route out of the Solar System, it has been possible to project its in-bound route as well.

Astronomers know that when our Solar System was forming, it effectively ejected comets and asteroids because of the orbits of the largest planets. Given that planetary systems are forming all around us, it stands to reason that other systems would also eject comets and asteroids, some of which would inevitably reach us.

One suggestion projects that Oumuamua heralded from the approximate start-point direction of the star Vega, in the constellation of Lyra. However, even averaging a speed of 95,000 km/h, it took so long for the object to make the trek to us that Vega was not near its current position when the asteroid was there. Computer simulations are at odds to those who back-projected Oumuamua as having come from that area.

Did Oumuamua come from a nearby star known as TYC4742-1027-1? While TYC4742-1027-1 could well have had its own Oort Cloud—a theoretical cloud or debris that can surround any star, first proposed by Dutch astronomer Jan Oort (1900–1992)—it may well have been that it just happened to be in the same vicinity as TYC4742-1027-1 and was not ejected from the cloud.

Other research suggests that the object came from the nearby "Pleiades Moving Group" of young stars, also known as the "Local Association." The Pleiades Moving Group is similar to the AB Doradus Moving Group, a group of about 30 associated stars that move through space together with the star AB Doradus. Groups of this nature are distinguished by their members, all of whom are around the same age, of the same composition, and therefore most likely formed in the same location. This group is the closest known moving group to Earth, with the Pleiades Moving Group possibly harboring objects like Oumuamua and carrying them many, many miles before ejecting them far away from the point that gravity first captured them.

The possibility exists that as our system has traveled through space, it has been brought into contact with a rich field of objects like Oumuamua, and that we should perhaps ready ourselves for an escalation of such asteroid sightings.

It has been speculated that Oumuamua originates from closer to home and is simply a Kuiper Belt object. For the object to have achieved such speed, fast enough to fling it around the Sun, it would have needed to be pushed by the gravity of an unknown planetary body, an encounter that perhaps occurred in the Kuiper Belt and projected Oumuamua towards the inner part of the Solar System. It could have sent the object in any direction—it just happened to be our direction in this instance.

The truth is that we remain unsure as to where the traveler came from, but one thing is almost certain: we won't see it again. But Oumuamua is not alone. Estimates suggest there could be as many as 10,000 similar objects "scouting" between the Sun and Neptune. Other estimates suggest a far greater figure, perhaps millions. The speed of the majority of these visitors is just too great, and most are simply not picked up. Possibly, one Oumuamua-sized object passes through the inner part of our Solar System every year, undetected! Even with the development of such high-powered telescopes as Pan-STARRS1, picking up and observing such an event remains only a chance occurrence.

Most importantly, Oumuamua and other objects could indeed carry the secrets of how ours and other solar systems formed, a vital key in understanding our very existence and place in the universe.

And Finally?

There can be no "finally" or "to sum up" with regard to the universe. Our power to see further, hear more clearly, and push back the boundaries continues to advance our knowledge. Yet, 96% of what's out there remains unseen, unheard, and beyond our comprehension.

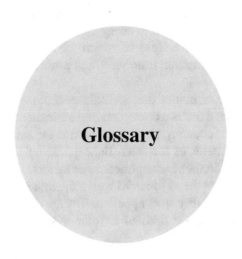

Glossary

ablation   A process whereby the atmosphere melts away the surface material of an incoming meteorite.

absolute magnitude   The brightness of a star or celestial object if seen from a standard distance of 10 parsecs.

achondrite   A stony meteorite lacking chondrules.

albedo   The ratio of the light reflected in all directions by a surface to the light incident on it. A perfectly reflecting surface has an albedo of 1; a perfectly absorbing surface has an albedo of 0.

altazimuth mount   The simplest type of telescopic mount with two motions, altitude (vertical) and azimuth (horizontal).

altitude   The angular distance between the direction to an object and the horizon. Altitude ranges from 0° for an object on the horizon to 90° for an object directly overhead.

amino acid   A carbon-based molecule from which protein molecules are assembled.

Amor asteroid   A member of a class of asteroids whose orbits cross the orbital distance of the Earth.

angular momentum   The momentum of a body associated with its rotation or revolution. For a body in a circular orbit, angular momentum is the product of orbital distance, orbital speed, and mass. When two bodies collide or interact, angular momentum is conserved.

© Springer Nature Switzerland AG 2018

J. Powell, *Rare Astronomical Sights and Sounds*, The Patrick Moore Practical Astronomy Series, https://doi.org/10.1007/978-3-319-97701-0

annihilation   The mutual destruction of a matter-antimatter pair of particles. The charges on the two particles cancel out, and the mass of the particles is entirely converted to energy.

annular eclipse   A solar eclipse in which the Moon is too far from the Earth to block the entire Sun from view, and a thin ring of sunlight appears around the Moon.

antimatter   A type of matter that annihilates ordinary matter on contact. For every particle, there is a corresponding antimatter particle. For example, the antimatter counterpart of the proton is the antiproton.

aperture   The diameter of the main light-gathering lens or mirror in an instrument, given in inches, centimeters, or meters.

apex   The direction in the sky toward which the Sun is moving. Because of the Sun's motion, nearby stars appear to diverge from the apex.

aphelion   The point in the orbit of a Solar System body that is farthest from the Sun.

apogee   The point, in an orbit about the Earth that is farthest from the Earth.

Apollo asteroid   A member of a class of asteroids whose orbits cross the orbital distance of the Earth.

apparent magnitude   The brightness of a star or celestial object when observed at its great distance from Earth.

asteroid   A small, planet-like solar system body. Most asteroids are rocky in makeup and have orbits of low eccentricity and inclination.

asteroid belt   The region of the Solar System lying between 2.1 and 3.3 astronomical units (AU) from the Sun. The great majority of asteroids are found in the asteroid belt.

astronomical unit (AU)   The mean earth-sun distance, about 150,000,000 km.

Aten asteroid   An asteroid having an orbit with a semi-major axis smaller than 1 AU.

atom   A particle consisting of a nucleus and one or more surrounding electrons.

atomic number   The number of protons in the nucleus of an atom. Unless the atom is ionized, the atomic number is also the number of electrons orbiting the nucleus of the atom.

Aurora Australis   Light emitted by atoms and ions in the upper atmosphere near the south magnetic pole. The emission occurs when atoms and ions are struck by energetic particles from the Sun.

Aurora Borealis   Light emitted by atoms and ions in the upper atmosphere near the north magnetic pole. The emission occurs when atoms and ions are struck by energetic particles from the Sun.

axis   The imaginary line that an object, usually a planet, rotates around.

azimuth   The angular distance between the north point on the horizon eastward around the horizon to the point on the horizon nearest to the direction of a celestial body.

Bailey's beads   Sunlight glimpsed through surface features around the limb of the Moon during a total solar eclipse. Also known as the "Diamond Ring" effect.

barred spiral galaxy   A spiral galaxy in which the nucleus is crossed by a bar. The spiral arms start at the ends of the bar.

barycenter   The center of mass of a system of bodies.

basalt   An igneous rock often produced in volcanic eruptions.

Big Bang   The theory that suggests that the Universe was formed from a single point in space during a cataclysmic explosion.

Big Crunch   The theory that states that the Universe will expand to its maximum point, then contract until it explodes.

Big Rip   A theory that states that all matter under the continued expansion of the Universe will eventually rip the Universe apart.

black hole   A region of space from which no matter or light can escape. A black hole is a result of the extreme curvature of space by a massive compact body.

bolide   A term used to describe an exceptionally bright meteor, possibly accompanied by a sonic boom.

bow shock   The region where the solar wind is slowed as it impinges on the Earth's magnetosphere.

brightness   Intensity of light received from a celestial object by the observer.

C-type asteroid   One of a class of very dark asteroids whose reflectance spectra show no absorption features due to the presence of minerals.

Capture Theory   The theory of the origin of the Moon, which holds that the Moon formed elsewhere in the Solar System and then was captured into orbit about the Earth.

carbonaceous chondrite   A stony meteorite that contains carbon-rich material. Thought to be samples of primitive material from the early Solar System.

Cassini's division   A conspicuous 1,800-km-wide gap between the outer-most rings of Saturn.

celestial equator   The circle where the Earth's equator, if extended outward into space, would intersect the celestial sphere.

celestial horizon   The circle on the celestial sphere which is 90° from the zenith. The celestial horizon is the approximate boundary between the Earth and sky.

celestial mechanics   The part of physics and astronomy that deals with the motions of celestial bodies under the influence of their mutual gravitational attraction.

celestial pole   The celestial poles are imaginary lines that trace the Earth's rotation axis in space.

celestial sphere   An imaginary sphere surrounding the Earth. The celestial bodies appear to carry out their motions on the celestial sphere.

centaurs   Small astronomical bodies that generally orbit the Sun between Jupiter and Neptune. Centaurs cross the orbital paths of the major planets.

charge coupled device (CCD)   An array of photosensitive electronic elements that can be used to record an image falling upon it. CCD cameras are composed of silicon chips that are light sensitive, changing detected photons of light into electronic signals that in turn can be used to create images of astronomical objects.

chondrite   A meteorite containing chondrules.

chondrule   A small, spherical body embedded in a meteorite. Chondrules are composed of iron, aluminium, and magnesium silicate rock.

chromosphere   The part of the Sun's atmosphere between the photosphere and the corona.

circumpolar stars   Stars that never set or go below the horizon for observers at specific latitudes.

close pair   A binary system in which the two stars are close enough together that they transfer matter to one another during some stages of their evolution.

cluster of galaxies   A group of galaxies held together by their mutual gravitational attraction.

cluster of stars   A group of stars held together by their mutual gravitational attraction.

coma   A spherical gaseous region that surrounds the nucleus of a comet. The coma of a comet may be 100,000 km or more in diameter.

comet   A small, icy body in orbit about the Sun. When a comet is near the Sun, it displays a coma and a tail.

concretions   A common geological phenomenon where hard bodies form in sediments before they become sedimentary rocks.

conjunction   The appearance of two celestial bodies, often a planet and the Sun, in approximately the same direction.

constellation   One of 88 regions into which the celestial sphere is divided.

continuous spectrum   A spectrum containing neither emission nor absorption lines.

convection   The process of energy transport in which heat is carried by hot and rising, and cool and falling, currents or bubbles of liquid or gas.

core   The innermost region of the interior of the Earth or another planet.

Coriolis effect   The acceleration that a body experiences when it moves across the surface of a rotating body. The acceleration results in a westward deflection of projectiles and currents of air or water when they move toward the Earth's equator, and an eastward deflection when they move away from the equator.

corona   The outermost layer of the Sun's atmosphere. Gases in the corona are tenuous and hot.

coronal hole   A dim, low-density region in the Sun's corona. Occur in regions of open magnetic field lines, where gases can flow freely away from the Sun to form the solar wind.

coronal mass ejection   A blast of gas moving outward through the Sun's corona and into interplanetary space following the eruption of a prominence.

cosmic background radiation (CBR)   Radiation observed to have almost perfectly uniform brightness in all directions in the sky. The CBR is highly redshifted radiation produced about a million years after the Universe began to expand.

cosmic ray   Extremely energetic ions and electrons that travel through space almost at the speed of light. Most cosmic rays come from great distances and may be produced in supernovas and pulsars

cosmic string   A tube-like configuration of energy that is believed to have existed in the early universe.

cosmology   The study of the universe as a whole.

crater   A roughly circular feature on the surface of a solar system body caused by the impact of an asteroid or comet.

crater density   The number of craters of a given size per unit area of the surface of a solar system body.

crescent phase   The phase of the Moon at which only a small, crescent-shaped portion of the near side of the Moon is illuminated by sunlight. Occurs just before and after new moon.

critical density   The value that the average density of the universe must equal or exceed if the universe is closed. If the density of the universe is less than the critical density, the universe will continue to expand forever.

crust   The outermost layer of the interior of a planet or satellite.

dark matter   Matter that cannot be detected or has not yet been detected by the radiation it emits. The presence of dark matter can be deduced from its gravitational interaction with other bodies.

dark energy   A theoretical form of energy postulated to act in opposition to gravity.

declination The angular distance of a celestial body north or south of the celestial equator. Declination is analogous to latitude in the terrestrial coordinate system.

degree A unit used to measured angles. There are 360 degrees in a circle.

density The mass of a body divided by its volume.

differential rotation When the rotation period of a body varies with latitude. Occurs for gaseous bodies like the Sun or for planets with thick atmospheres.

differentiation The gravitational separation of the interior of a planet into layers according to density. When differentiation occurs inside a molten body, the heavier materials sink to the center and the light materials rise to the surface.

doppler effect The change in the frequency of a wave (such as electromagnetic radiation) caused by the motion of the source and observer toward or away from one another.

dust tail A comet tail that is luminous because it contains dust that reflects sunlight. The dust in a comet tail is expelled from the nucleus of the comet.

earthshine A dull glow that lights up the unlit part of the Moon because the Sun's light reflects off the Earth's surface and back onto the Moon.

Ebbinghaus illusion An optical illusion of relative size perception.

eclipse The obscuration of the light from the Sun when the observer enters the Moon's shadow or the Moon when it enters the Earth's shadow. Also, the obscuration of a star when it passes behind its binary companion.

ecliptic The plane of the Earth's orbit about the Sun. As a result of the Earth's motion, the Sun appears to move among the stars, following a path that is also called the ecliptic.

electromagnetic wave A periodic electrical and magnetic disturbance that propagates through space and transparent materials at the speed of light. Light is an example of an electromagnetic wave.

electron A low-mass, negatively charged particle that can either orbit a nucleus as part of an atom or exist independently as part of a plasma.

element A substance that cannot be broken down into a simpler chemical substance. Oxygen, nitrogen, and silicon are examples of the approximately 100 known elements.

ellipse A closed, elongated curve describing the shape of the orbit that one body follows about another.

elliptical galaxy A galaxy having an ellipsoidal shape and lacking spiral arms.

elongation The Angular distance of a celestial object from the sun in the sky.

ephemeris   A tabulation of the positions of a celestial object in sequence for a succession of dates.

equator   The line around the surface of a rotating body that is midway between the rotational poles. The equator divides the body into northern and southern hemispheres.

equatorial jet   The high-speed, eastward, zonal wind in the equatorial region of Jupiter's atmosphere.

equatorial mount   A telescope mount that allows the observer to follow the rotation of the sky as the Earth turns.

equatorial system   A coordinate system using right ascension and declination as coordinates, used to describe the angular location of bodies in the sky.

equinox   Either of the two points on the celestial sphere where the ecliptic intersects the celestial equator.

escape velocity   The speed required by an object to achieve a parabolic trajectory and escape from its parent body.

event horizon   The boundary of a black hole. No matter or radiation can escape from within the event horizon.

exosphere   The outer part of the thermosphere. Atoms and ions can escape from the exosphere directly into space.

eyepiece   The lens at the viewing end of a telescope.

Fermi paradox   The question: Given the known size of the Universe, why have we not been contacted and are still alone? Named after Italian physicist Enrico Fermi (1901-1954).

filament   A dark line on the Sun's surface when a prominence is seen projected against the solar disk.

fireball   An especially bright streak of light in the sky produced when an interplanetary dust particle enters the Earth's atmosphere, vaporizing the particle and heating the atmosphere.

focal length   The distance between a mirror or lens and the point at which the lens or mirror brings light to a focus.

focal plane   The surface where the objective lens or mirror of a telescope forms the image of an extended object.

focal point   The spot where parallel beams of light striking a lens or mirror are brought to a focus.

fusion   A nuclear reaction in which two nuclei merge to form a more massive nucleus.

galactic bulge   A somewhat flattened distribution of stars surrounding the nucleus of the Milky Way.

galactic disk   A disk of matter containing most of the stars and interstellar matter in the Milky Way.

galactic equator The great circle around the sky that corresponds approximately to the center of the glowing band of the Milky Way.

galactic halo The roughly spherical outermost component of the Milky Way.

galactic nucleus The central region of the Milky Way.

galaxy A massive system of stars, gas, and dark matter held together by its own gravity.

gamma ray The part of the electromagnetic spectrum having the shortest wavelengths.

gegenschein A patch of very faint nebulous light sometimes seen in the night sky opposite the position of the Sun.

geosynchronous orbit An orbit in which a satellite's orbital velocity is matched to the rotational velocity of the planet.

globular cluster A tightly packed, spherically shaped group of thousands to millions of old stars.

granule A bright convective cell or current of gas in the Sun's photosphere. Granules appear bright because they are hotter than the descending gas that separates them.

gravitational lens A massive body that bends light passing near it. Can distort or focus the light of background sources of electromagnetic radiation.

gravity The force of attraction between two bodies generated by their masses.

great attractor A great concentration of mass towards which everything in our part of the universe apparently is being pulled.

greenhouse effect The blocking of infrared radiation by a planet's atmospheric gases. Because its atmosphere blocks the outward passage of infrared radiation emitted by the ground and lower atmosphere, the planet cannot cool itself effectively and becomes hotter than it would be without an atmosphere.

habitable zone The range of distances from a star within which liquid water can exist on the surface of an Earth-like planet.

helioseismology A technique used to study the internal structure of the Sun by measuring and analysing oscillations of the Sun's surface layers.

heliosphere The region of space dominated by the solar wind and the Sun's magnetic field.

Hilda asteroids A group of asteroids with a 3:2 orbital resonance with Jupiter.

Hubble's law The linear relationship between the recession speeds of galaxies and their distances. The slope of Hubble's law is Hubble's constant.

hyperbola A curved path that does not close on itself. A body moving with a speed greater than escape velocity follows a hyperbola.

igneous rock   A rock formed by solidification of molten material.

inclination   The tilt of the rotation axis or orbital plane of a body.

inertia   The tendency of a body at rest to remain at rest and a body in motion to remain in motion at a constant speed and in constant direction.

inertial motion   Motion in a straight line at constant speed when there are no unbalanced forces acting upon the body.

inferior planet   A planet whose orbit lies inside the Earth's orbit.

infrared   The part of the electromagnetic spectrum having wavelengths longer than visible light but shorter than radio waves.

interferometry   The use of two or more telescopes connected together to operate as a single instrument. Interferometers can achieve high angular resolution if the individual telescopes of which they are made are widely separated.

interstellar matter   Gas and dust in the space between the stars.

ion   An atom from which one or more electrons has been removed.

ionization   The removal of one or more electrons from an atom.

ionosphere   The lower part of the thermosphere of a planet in which many atoms have been ionized by ultraviolet solar photons.

inferior conjunction   A conjunction of an inferior planet that occurs when the planet is lined up directly between the Earth and the Sun.

iron meteorite   A meteorite composed primarily of iron and nickel.

isotopes   Nuclei with the same number of protons but different numbers of neutrons.

jets (comets & galaxies)   Venting of gas from weaken areas of a comet's nucleus. Also, a narrow beam of gas ejected from a star or the nucleus of an active galaxy.

Kardashev scale   A method of measuring a civilization's level of technological advancement, formulated by Russian astronomer Nikolai Kardashev.

Kepler's laws of planetary motion   Three laws, discovered by Kepler, that describe the motions of the planets around the Sun

kinetic energy   Energy of motion. Kinetic energy is given by one half the product of a body's mass and the square of its speed.

Kirkwood gaps   Regions in the asteroid belt where a decreased number of asteroids are found, possibly the result of believed to be the result of gravitational interactions with Jupiter. Named after astronomer Daniel Kirkwood (1814-1895), who first observed these gaps.

Kuiper belt   A region beyond Neptune, within which a large number of comets are believed to orbit the Sun. Short-period comets are thought to originated in the Kuiper belt.

Lagrangian points   Positions in an orbital configuration where a small body, under the gravitational influence of two larger ones, will remain approximately at rest relative to them. Named after 18th century Italian astronomer and mathematician Joseph-Louis Lagrange, (1736-1813).

latitude   The angular distance of a point north or south of the equator of a body as measured by a hypothetical observer at the center of a body.

lava   Molten rock at the surface of a planet or satellite.

libration points   (See Lagrangian Points).

light   The visible form of electromagnetic radiation.

light curve   A plot of the brightness of a body versus time.

light year (ly)   A unit of length equal to the distance that light travels in one year in a vacuum, about 9.46 trillion km.

limb   The apparent edge of the disk of a celestial body.

lithosphere   The rigid outer layer of a planet or satellite, composed of the crust and upper mantle.

Local Group   A small, specific cluster of galaxies, of which the Milky Way is a member.

longitude   The angular distance around the equator of a body from a zero point to the place on the equator nearest a particular point, as measured by a hypothetical observer at the center of a body.

luminosity   The rate of total radiant energy output of a body.

luminosity class   The classification of a star's spectrum according to luminosity for a given spectral type. Luminosity class ranges from "I" for a supergiant to "V" for a dwarf (main sequence star).

luminosity function   The distribution of stars or galaxies according to their luminosities. A luminosity function is often expressed as the number of objects per unit volume of space that are brighter than a given absolute magnitude or luminosity.

lunar eclipse   The darkening of the Moon that occurs when the Moon enters the Earth's shadow.

lunation   Also known as a synodic month, it is the average time from one new Moon to the next.

M-type asteroid   One of a class of asteroids that have reflectance spectra like those of metallic iron and nickel.

Magellanic Clouds   Two irregular galaxies that are among the nearest neighbors of the Milky Way.

magma   Molten rock within a planet or satellite.

magnetosphere   The outermost part of the atmosphere of a planet, within which very thin plasma is dominated by the planet's magnetic field.

magnitude   A number, based on a logarithmic scale, used to describe the brightness of a star or other luminous body. "Apparent magnitude" describes the brightness of a star as we see it. "Absolute magnitude" describes the intrinsic brightness of a star.

mantle   The part of a planet lying between its crust and its core.

maria   A dark, smooth region on the Moon formed by flows of basaltic lava.

mass   A measure of the amount of matter a body contains. Mass is also a measure of the inertia of a body.

maunder minimum   A period of few sunspots and low solar activity that occurred between 1640 and 1700.

mean solar time   Time kept according to the average length of the solar day.

meridian   The circle on the celestial sphere that passes through the zenith and both celestial poles.

mesosphere   The layer of a planet's atmosphere above the stratosphere. The mesosphere is heated by absorbing solar radiation.

Messier objects   A famous list of deep sky objects compiled by Charles Messier (1730-1817).

metallic hydrogen   A form of hydrogen in which the atoms have been forced into a lattice structure typical of metals. In the Solar System, the pressures and temperatures required for metallic hydrogen to exist only occur in the cores of Jupiter and Saturn.

metamorphic rock   A rock that has been altered by heat and pressure.

meteor   A streak of light produced by meteoroid moving rapidly through the Earth's atmosphere. Friction vaporizes the meteoroid and heats atmospheric gases along the path of the meteoroid.

meteor shower   A temporary increase in the normal rate at which meteors occur. Meteor showers last for a few hours or days and occur on about the same date each year.

meteorite   The portion of a meteoroid that reaches the Earth's surface.

meteoroid   A solid interplanetary particle passing through the Earth's atmosphere.

microlensing event   The temporary brightening of a distant object that occurs because its light is focused on the Earth by the gravitational lensing of a nearer body.

micrometeorite   A meteoritic particle less than a 50 millionths of a meter in diameter. Micrometeorites are slowed by atmospheric gas before they can be vaporized, so they drift slowly to the ground.

Milky Way   The galaxy to which the Sun and Earth belong. Seen as a pale, glowing band across the sky.

mineral   A solid chemical compound.

minor planet   Another name for asteroid.

molecular cloud   A relatively dense, cool interstellar cloud in which molecules are common.

momentum   A quantity equal to the product of a body's mass and velocity, used to describe the motion of the body. When two bodies collide or otherwise interact, the sum of their momenta is conserved.

near Earth asteroid (NEAR) or near Earth object (NEO)   Bodies who orbits come into close proximity with Earth.

neutral gas   A gas containing atoms and molecules but essentially no ions or free electrons.

neutrino   A particle with no charge and probably no mass that is produced in nuclear reactions. Neutrinos pass freely through matter and travel at or near the speed of light.

neutron   A nuclear particle with no electric charge.

neutron star   A star composed primarily of neutrons and supported by the degenerate pressure of the neutrons.

neutronization   A process by which, during the collapse of the core of a star, protons and electrons are forced together to make neutrons.

noctilucent clouds   A cloud-like phenomenon in the upper atmosphere of the Earth caused by the presence of ice crystals. Also known as "night shining."

north celestial pole   The point above the Earth's north pole where the Earth's polar axis, if extended outward into space, would intersect the celestial sphere.

nova   An explosion on the surface of a white dwarf star in which hydrogen is abruptly converted into helium.

nucleus   An irregularly shaped, loosely packed lump of dirty ice several km across that is the permanent part of a comet.

objective   The main lens or mirror of a telescope.

occultation   An event that occurs when one celestial body conceals or obscures another.

Oort cloud   The region beyond the planetary system, extending to 100,000 AU or more, within which a vast number of comets orbit the Sun. When comets from the Oort cloud enter the inner Solar System, they become new comets.

opposition   The configuration of a planet or other body when it appears opposite the Sun in the sky.

orbit   The elliptical or circular path followed by a body that is bound to another body by their mutual gravitational attraction.

organic molecule   A molecule containing carbon.

oscillating universe   A theory that the universe goes through continual phases of expansion and contraction.

outgassing   The release of gas from the interior of a planet or satellite.

ozone   A molecule consisting of three oxygen atoms. Ozone molecules are responsible for the absorption of solar ultraviolet radiation in the Earth's atmosphere.

parabola   A geometric curve followed by a body that moves with a speed exactly equal to escape velocity.

parallax   The shift in the direction of a star caused by the change in the position of the Earth as it moves about the Sun.

parsec   A unit of distance equal to about 3.26 ly.

penumbra   The outer part of the shadow of a body where sunlight is partially blocked by the body.

perigee   The point in an orbit about the Earth that is nearest to the Earth.

perihelion   The point in the orbit of a body when it is closest to the Sun.

perturbation   A deviation of the orbit of a solar system body from a perfect ellipse due to the gravitational attraction of one of the planets.

photon   A massless particle of electromagnetic energy.

photometry   The measurement of the light emitting from astronomical objects.

photosphere   The visible region of the atmosphere of the Sun or another star.

planetesimal   A primordial solar system body of intermediate size that accreted with other planetesimals to form planets and satellite.

plasma   A fully or partially ionized gas.

plasma tail   A narrow, ionized comet tail pointing directly away from the Sun.

plate tectonics   Describes the movement of seven large plates and a number of smaller plates of the Earth's lithosphere.

potentially hazardous asteroid (PHA)   A group of asteroids that carry a collision potential with Earth.

precession   The slow, periodic conical motion of the rotation axis of the Earth or another rotating body.

prominence   A region of cool gas embedded in the corona. Prominences are bright when seen above the Sun's limb, but appear as dark filaments when seen against the Sun's disk.

proper motion   The rate at which a star appears to move across the celestial sphere with respect to very distant objects.

protein   A large molecule consisting of a chain of amino acids, which makes up the bodies of organisms.

proton   A positively charged nuclear particle.

protostar   A star in the process of formation.

pulsar   A rotating neutron star with beams of radiation emerging from its magnetic poles. When the beams sweep past the Earth, we see "pulses" of radiation.

quantum mechanics   The branch of physics dealing with the structure and behaviour of atoms and their interaction with light.

quasar   A distant galaxy, seen as it was in the remote past, with a very small, luminous nucleus.

radial velocity   The part of the velocity of a body that is directed toward or away from an observer. The radial velocity of a body can be determined by the Doppler shift of its spectral lines.

radiant   The point in the sky from which the meteors in a meteor shower seem to originate.

radio galaxy   A galaxy that is a strong source of radio radiation.

radioactivity   The spontaneous disintegration of an unstable nucleus of an atom.

reflectivity   The ability of a surface to reflect electromagnetic waves. The reflectivity of a surface ranges from 0% for a surface that reflects no light to 100% for a surface that reflects all the light falling on it.

reflector   A telescope in which the objective is a mirror.

refractor   A telescope in which the objective is a lens.

regolith   The surface layer of dust and fragmented rock, caused by meteoritic impacts, on a planet, satellite, or asteroid.

resolution   The ability of a telescope to distinguish fine details of an image.

resonance   The repetitive gravitational tug of one body on another when the orbital period of one is a multiple of the orbital period of the other.

retrograde motion   The westward revolution of a Solar System body around the Sun.

right ascension   Angular distance of a body along the celestial equator from the vernal equinox eastward to the point on the equator nearest the body. Right ascension is analogous to longitude in the terrestrial coordinate system.

Roche limit or Roche radius   The distance from a planet or other celestial body within which tidal forces from the body would disintegrate a smaller object. Term formulated by French mathematician Édouard Roche (1820-1833).

S-type asteroid   One of a class of asteroids whose reflectance spectra show an absorption feature due to the mineral olivine.

Search for Extra-Terrestrial Intelligence (SETI)   A NASA-led project to search for extra-terrestrial intelligence.

sedimentary rock   A rock formed by the accumulation of small mineral grains carried by wind, water, or ice to the spot where they were deposited.

seismic wave   Waves that travel through the interior of a planet or satellite and are produced by earthquakes or their equivalent.

sidereal clock   A clock that marks the local hour angle of the vernal equinox.

silicate   A mineral whose crystalline structure is dominated by silicon and oxygen atoms.

solar constant   The solar energy received by a square meter of surface oriented at right angles to the direction to the Sun at the Earth's average distance (1 AU) from the Sun. The value of the solar constant is 1,372 watts per square meter.

solar flare   A brief, sudden brightening of a region of the Sun's atmosphere, probably caused by the abrupt release of magnetic energy.

solar maximum   The period of greatest solar activity in the 11-year solar cycle.

solar minimum   The period of least solar activity in the 11-year solar cycle.

spectral class   A categorization based on the pattern of spectral lines of stars that groups stars according to their surface temperatures.

spectrograph   A device used to produce and record a spectrum.

spectroscopy   The recording and analysis of spectra.

spicule   A hot jet of gas moving outward through the Sun's chromosphere.

spiral arm   A long, narrow feature of a spiral galaxy in which interstellar gas, young stars, and other young objects are found.

spiral galaxy   A flattened galaxy in which hot stars, interstellar clouds, and other young objects form a spiral pattern.

star   A massive gaseous body that has used, is using, or will use nuclear fusion to produce the bulk of the energy it radiates into space.

starburst galaxy   A galaxy in which a very large number of stars have recently formed.

steady state theory   A cosmological theory in which the U = universe always remains the same in its essential features, such as average density. In order to maintain constant density while expanding, the steady state theory required the continual creation of new matter.

stellar occultation   The obstruction of the light from a star when a solar system body passes between the star and the observer.

stellar parallax   The shift in the direction of a star caused by the change in the position of the Earth as it moves about the Sun.

stellar population   A group of stars similar in spatial distribution, chemical composition, and age.

stony meteorite   A meteorite made of silicate rock.

stony-iron meteorite   A meteorite made partially of stone and partially of iron and other metals.

stratosphere   The region of the atmosphere of a planet immediately above the troposphere.

sublimation   The change of a solid directly into a gaseous state.

sunspot A region of the Sun's photosphere that appears darker than its surroundings because it is cooler.

sundog An atmospheric optical phenomenon that consists of a bright spot to the left or right of the Sun, or equally together.

sunspot cycle The regular waxing and waning of the number of spots on the Sun. The amount of time between one sunspot maximum and the next is about 11 years.

sun pillar An atmospheric optical phenomenon whereby a shaft of light is seen extending upward or downward from the Sun.

sunspot group A cluster of sunspots.

superior conjunction A conjunction that occurs when a planet passes behind the Sun and is on the opposite side of the Sun from the Earth.

supernova An explosion in which a star's brightness temporarily increases by as much as 1 billion times. Type I supernovas are caused by the rapid fusion of carbon and oxygen within a white dwarf. Type II supernovas are produced by the collapse of the core of a star.

synchronous rotation Rotation for which the period of rotation is equal to the period of revolution. An example of synchronous rotation is the Moon, for which the period of rotation and the period of revolution about the Earth are both one month.

synodic month The length of time (29.53 days) between successive occurrences of the same phase of the Moon (see lunation).

synodic period The length of time it takes a Solar System body to return to the same configuration (opposition to opposition, for example) with respect to the Earth and the Sun.

tektite A small, glassy material formed by the impact of a large body, usually a meteor or asteroid.

terminal velocity The speed with which a body falls through the atmosphere of a planet when the force of gravity pulling it downward is balanced by the force of air resistance.

thermosphere The layer of the atmosphere of a planet lying above the mesosphere. The lower thermosphere is the ionosphere. The upper thermosphere is the exosphere.

transverse velocity The part of the orbital speed of a body perpendicular to the Sun between the body and the Sun.

Trojan asteroid One of a group of asteroids that orbit the Sun at Jupiter's distance and lie 60° ahead of or behind Jupiter in its orbit.

troposphere The lowest layer of the atmosphere of a planet, within which convection produces weather.

ultraviolet The part of the electromagnetic spectrum with wavelengths longer than X-rays but shorter than visible light.

umbra   The inner portion of the shadow of a body, within which sunlight is completely blocked.

V-type asteroid   The asteroid Vesta, which is unique in having a reflectance spectra resembling those of basaltic lava flows.

Van Allen belts   Two doughnut-shaped regions in the Earth's magnetosphere within which many energetic ions and electrons are trapped.

velocity   A physical quantity that gives the speed of a body and the direction in which it is moving.

visual binary star   A pair of stars orbiting a common center of mass, in which the images of the components can be distinguished using a telescope and which have detectable orbital motion.

wavelength   The distance between crests of a wave. For visible light, wavelength determines color.

weakly interacting massive particles (WIMPS)   WIMPS have 10 to 100 times the mass of a proton.

white hole   A hypothetical region of spacetime that cannot be entered from the outside, although matter and light can escape from it. The reverse of a black hole.

wormhole   A speculative feature of a black hole that proposedly connects our universe with another universe.

X-ray   The part of the electromagnetic spectrum with wavelengths longer than gamma rays but shorter than ultraviolet.

X-ray burst   A sporadic burst of X-rays originating in the rapid consumption of nuclear fuels on the surface of the neutron star in a binary system.

zenith   The point on the celestial sphere directly above an observer.

zodiacal constellations   The band of constellations along the ecliptic. The Sun appears to move through the 12 zodiacal constellations during a year

zodiacal light   The faint glow extending away from the Sun, caused by the scattering of sunlight by interplanetary dust particles lying in and near the ecliptic

zonal winds   The pattern of winds in the atmosphere of a planet in which the pattern of wind speeds varies with latitude.

zone of convergence   According to plate tectonics, a plate boundary at which the crustal plates of a planet are moving toward one another. Crust is destroyed in zones of convergence.

zone of divergence   According to plate tectonics, a plate boundary at which the crustal plates of a planet are moving away from one another. Crust is created in zones of divergence.

Index

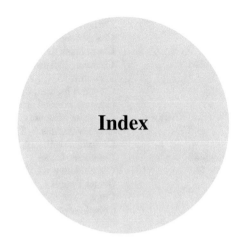

© Springer Nature Switzerland AG 2018

J. Powell, *Rare Astronomical Sights and Sounds*, The Patrick Moore
Practical Astronomy Series, https://doi.org/10.1007/978-3-319-97701-0

213